[re]wired with AI

Selling your AE services in an AI-enabled world

[re]wired with AI

Selling your AE services
in an AI-enabled world

by

David A. Stone

with assistance from
AI-powered collaborators

WINDWORD GROUP
PUBLISHING & MEDIA

Published in the United States by
Windword Group Publishing & Media
2950 Newmarket Street, Suite 101-297
Bellingham, WA 98226
info@windwordgroup.com.

ISBN 978-1-947527-63-8 (paperback)

DISCLAIMER

The information in this book is provided for general educational and professional development purposes. It is not intended, and should not be relied upon, as legal, contractual, or regulatory advice. Readers should consult qualified legal counsel or compliance professionals for guidance specific to their organization, jurisdiction, or client agreements.

Note on Language:

Throughout this book we use the term **AI assistant** to refer broadly to conversational or generative-AI tools such as ChatGPT, Claude, or other enterprise-approved systems. "Assistant" reminds us that these tools *support* human expertise. They don't replace it.

Note on Authorship

This book was developed with the support of generative-AI writing systems. All final interpretations, edits, and ideas are the work of the author.

ALSO BY DAVID A. STONE

[re]wired: Selling your AE services in a post-recession world (with Gail Hulnick)

Nothing Happens Till Somebody Sells Something

Nothing Happens Till Somebody Sells Something 2

Nothing Happens Till Somebody Sells Something 3 (with Gail Hulnick)

What I Need to Know About Marketing

TABLE OF CONTENTS

PREFACE

In 2015 I collaborated with Gail Hulnick to write a book titled, *[re]wired: Selling your AE services in a post-recession world.* In that book, we invested a lot of energy and words in an effort to show you that, when it comes to selling your professional design services, it isn't enough to simply convey "the facts" about your firm and let the client conclude that you're the best choice.

If you haven't read the previous book, read it now. It's still enormously relevant.

By the time that book was written, competition among AE firms had already become intense. In the chase after any given project, every firm hoped to show that they were, in the words of the Brooks Act as it espoused QBS, "the most qualified." The challenge had become that every firm was fully qualified to do the work. The truly incompetent had been weeded out and the firms remaining in the hunt—often as many as 50, competing for the same project—were competent, even beyond competent. How do you select the "most" qualified from a large group that are all "fully" qualified?

In the first book we referred to what we called, the "art" of persuasion; the ability to get another person to see what you see, to understand and believe what you do. Marketing and sales were, are still, and will always be about human-to-human persuasion. That skill has never involved throwing page after page of project lists at a prospective client. It has always, and will always be about empathy and your ability to understand their needs, their fears, their desires, and their unique view of the world.

Since that first book was written, thousands of firms have adopted the techniques we suggested in it. The quality of marketing materials, proposals, and presentations has improved dramatically, and the bar of client expectations has been raised accordingly. Keep up the good work!

But one dramatic and earth-shattering change has occurred since then—the introduction and widespread access to Artificial Intelligence.

Various pundits have referred to this change as being more revolutionary than the printing press, the industrial revolution, or even the invention of the silicone chip. History will tell.

In the meantime, AI is a fact with which you must now come to grips and I want to be clear: This book isn't about whether you should or shouldn't adopt AI in your marketing and sales. You should and you must. This book is about how.

According to the Pew Research Center, as of this writing (Spring, 2026) "roughly half of U.S. adults (52%) feel more concerned than excited about the increased use of AI in daily life. Anxiety is high regarding job security, with ~65–75% of employees fearing AI will make their jobs obsolete or negatively affect their career, while 88% of non-users are unclear about AI's impact on their lives."

If you are among the anxious majority, know this: AI has already changed the way you work and will continue to. But, as with the introduction of any disruptive technology, those who understand and embrace the change are the ones who are positioned to lead and benefit from it.

Unlike any of my previous twenty-something books, all of which I wrote the old-fashioned way, I have written this one with the active collaboration of AI Assistants. It has been a fascinating act of congruence: to collaborate with the very tools I'm discussing.

The process has been that I would lay out my intentions and AI would offer draft beats—portions of chapters—in response. I would then comment and provide additional direction, to which it would return more fully fleshed-out passages. The iterative process continued, sometimes at length, until I was satisfied with the gist of the material. Then I edited, contextualized, and humanized the material to arrive at the final work.

In a way, it has been a grand experiment on my part. Now well into my eighth decade of dealing with change, I genuinely did not know, heading into this project, whether it would work or not. Nor did I know how I would feel, working with a robot.

The conclusion is that I've loved every minute of it. And had my mind expanded at a breathtaking pace. The partnership has felt exactly like the premise of this book—human judgment leading, machine intelligence supporting. The spark is still created by conversation, not automation.

So, if you hold this book looking for the next instruction manual or the next "system" to install, you might miss its larger heartbeat. This is about you and the newly empowered ways in which you can do what you've always done. You don't need, and you won't find a replacement for your expertise. You've merely supercharged it.

David Stone
April, 2026

INTRODUCTION

The Momentum Continues: Why This Isn't a "2nd Edition"

When *[re]wired: Selling your services in a post-recession world* first came out in 2015, the world of professional services was still clawing its way out of economic contraction. Firms were scrambling to re-win work, rebuild client confidence, and reimagine how to market in a world that suddenly felt smaller, faster, and more risk-averse. That book captured a moment — a time when "selling your AE services in a post-recession world" meant rethinking relationships, proposals, and value from the ground up.

And it worked. Thousands of design professionals read it, argued with it, built their marketing processes from it, quoted it, or tucked it into the top drawer of their desk as a reminder that even in lean times, smart communication wins.

Then, quietly at first and then all at once, the ground shifted again.

This new book exists because the rules of the marketplace—the speed, the tools, the expectations—have changed dramatically. The people, however, have not.

Clients still choose based on trust, understanding, and emotional fit. They still buy certainty, simplicity, and confidence. They don't really want a new vendor; they want an old friend who's using new tools.

So, no, this isn't a "second edition." If the first *[re]wired* was a response to a financial recession, *[re]wired with AI* is a response to an attention recession. It's not about recovering lost ground.

It's about regaining focus, energy, and human presence in an age that's louder, faster, and algorithmically relentless.

Where the first book taught teams how to communicate more persuasively in a wary market, this one shows leaders how to stay human in a digital one.

A Living System, Not a Sequel

Every strong idea eventually meets a new condition that tests its flexibility. The principles in the original *[re]wired*—clarity, empathy, structure, story—remain rock solid. The idea of "wiring" your marketing and business-development strategy was about building circuits of trust and momentum through a team. That metaphor still holds.

But look at how the world has transformed since: smartphones became prosthetic brains, client relationships migrated to screens, and attention spans fragmented into notifications. Entire pursuits—strategy sessions, interviews, presentations—moved from conference rooms to video calls. By the time you've finished this paragraph, your phone might have vibrated twice. Everyone's has.

That's why this is not a revision; it's a rewiring for a new current. The conductors are the same; the voltage is completely different.

This book imagines what happens when AI enters that circuitry—not as a threat to your expertise, but as a booster to it; not as a shortcut, but as a force multiplier.

Why Now?

When we wrote the original book, most firms were just discovering digital marketing as something beyond a brochure. Websites were finally more than online business cards. A few intrepid marketers were dabbling in analytics and social media, experimenting with early CRM tools, trying to prove the worth of marketing data to skeptical principals.

Fast forward a decade, and entire disciplines such as data visualization, content strategy, SEO, and automation have become standard. We no longer debate whether technology belongs in business development; we debate how much is too much. And then, just as we begin to feel steady, AI explodes onto the scene.

In under two years, "prompt engineering" becomes a daily skill. Image generation, language models, and analytics co-pilots are everywhere, embedded invisibly in your email, your browser, and your CRM. Suddenly, professionals who never thought of themselves as "tech people" are carrying a small think-tank in their pocket. The tools have become accessible and astonishingly powerful.

Yet for all this progress, something subtle has frayed: the easy confidence of knowing when your message lands. Screens flatten nuance. Automation blurs authenticity. And the race to keep up can leave even the brightest professionals feeling behind before they've begun.

That's where this book steps in. It's here to rebuild calm, competence, and strategic clarity. I'm going to show you that AI doesn't replace the human advantage. It refocuses it.

Evolving the Original Promises

If you were part of the first *[re]wired* community, here's what hasn't changed:

- The focus on practicality. Every concept still resolves into something you can do tomorrow morning.
- The insistence that marketing and selling are team sports. Tools don't build culture. People do.
- The belief that clarity beats cleverness. The flashiest pitch still fails if the client can't see themselves in it.

But here's what is new:

- AE firms now sell and serve in parallel with machines that think. At least they convincingly simulate thinking.

- The members of professional design teams have skills that range from digital natives to digital skeptics.
- Firms operate in an economy where the difference between thriving and surviving often comes down to how quickly they can learn, adapt, and communicate across this mixed human-machine workspace.

This book meets that moment. It gives you a framework to evaluate new technology through the lens of human payoff. It asks of every shiny tool: Does this make us smarter? Faster? Kinder? Or just busier? Only the first three are worth the effort.

Continuing Momentum

Momentum is what the first book was really about. It encouraged that energy between people when ideas move forward. Then, the pandemic massively disrupted it and we're still trying to sort out the most effective to work in this hybrid environment. AI can help us find that balance if we channel it correctly. The opportunity now is enormous: to combine the unflappable dignity of expertise with the agility of constant learning.

Think of this as a field manual for optimism. It's not a book about technology; it's a book about staying relevant, valuable, and deeply human while leveraging the technology.

Whatever your role—principal, marketer, seller-doer, or emerging leader—you'll find that the chapters bring useful, actionable insight, information, strategies, and tactics when you have proposals to write, clients to keep, teams to motivate, and technology to evaluate.

How the world (and work) has changed since *[re]wired* was first published

When the original *[re]wired* landed, business development still smelled faintly of toner from printed proposals. Client meetings mostly happened face to face. LinkedIn was just warming up, Zoom was an optional convenience, and "virtual coffee" hadn't yet entered anyone's vocabulary. The practice of marketing and selling in the AE world revolved around relationships built in person—the handshake, the lunch, the site visit, the conference hallway.

Today, a decade later, those moments are still precious. But they live in a completely different ecosystem. Now, the meeting may be virtual, the rapport may start with pixels, and your best argument for credibility might be an invisible algorithm running beneath your email signature.

It's not that the fundamentals of persuasion changed; it's that the context around them has exploded in complexity.

The macro shift: From linear to lightning

When I began my career as an architect, the old rhythm of professional communication had clear phases. Research, write, review, present, wait. Days or weeks between each step. Now, the entire cycle compresses into hours.

Marketing once moved at the speed of print. Today it moves at the speed of "seen at 8:17 a.m., replied at 8:19." The expectation of instant response has infiltrated how clients perceive reliability. If you can't keep pace digitally, it reads as indifference, not restraint.

Meanwhile, the marketplace globalized. An architecture firm in Denver competes for the same client as one in Dublin. The first page of search results has replaced the local short list. Reputation is still everything but now it's packaged as discoverability.

Information, of course, is limitless. But attention is not. The challenge isn't reaching audiences; it's earning focus long enough for meaning to form. And the firms that once prided themselves on deliberate process now face a world that rewards strategic responsiveness. Never panic, but always motion.

This acceleration hasn't diminished the need for quality thinking; it's magnified it. When every competitor is shouting, silence can look like confidence, but only if it's paired with visible intelligence elsewhere. Every client is wondering and scanning: Who seems awake to the moment?

The emotional shift: From certainty to overload

Even before the pandemic, decision fatigue was becoming the new burnout. The sudden migration to remote work didn't just change where we sat; it changed how we processed. In a world of constant surveillance metrics, including pings, reminders, and unread counts, focus became fragile.

Our clients feel it too. They're not looking for more data. They're looking for meaning inside the data. They want fewer tabs open, not more. The seller who can take a torrent of information and distill it into one confident, human narrative immediately cuts through the noise.

This is one of the great modern ironies: the more connected we become; the lonelier business can feel. The more "accessible" everyone is, the harder it is to connect. Trust now forms differently than before. It's built through micro-moments on screens, through the tone of your writing, responsiveness, and small digital gestures that signal attentiveness.

Empathy has become a differentiator. Not the easy kind that lives in mission statements, but the disciplined kind that shows up when a partner says, "I read what you sent and here's what it means for you." That sentence, in 2026 and beyond, is gold.

As information multiplies, the market doesn't crave new voices; it craves translators. And that, incidentally, is what the best marketers and seller-doers have always been: interpreters between what clients desire and fear, and what firms can deliver.

The disruption: AI and the end of the blank page

As if we weren't jostled enough, along came the latest jolt: artificial intelligence entering everyday work. For years, we have treated "AI" as background noise in tech media, something futurists talked about at conferences. Then, almost overnight, it became a colleague. And sometimes an unnerving one.

Suddenly, the blank page isn't blank. The cursor can move itself. The barrier to "starting" evaporates. A tool can summarize transcripts, mine databases, generate concept imagery, or draft a first email faster than coffee can cool.

For creative professionals, this new competence can feel like cheating. For managers, it feels like opportunity wrapped in dizziness. For clients, it feels like magic they expect you to wield responsibly.

And yet the deeper you explore these systems, the clearer it becomes: AI doesn't invent originality; it mirrors it. It pattern-matches brilliance but doesn't possess intent. It's both humbling and liberating to realize that the most valuable work: curation, judgment, and care, is still stubbornly human.

The best leaders are already using AI not as an autopilot, but as an augmentation layer. They free their teams from early drafts and repetitive formatting so they can spend time on the insight, not the administration.

But here's the subtle danger: efficiency can quickly masquerade as progress. Producing more isn't the same as connecting more. Knowing when not to automate is now as strategic as knowing when to delegate. This book will help draw those lines.

New world, same wiring

So here we are: a profession of thinkers, creators, and problem-solvers, navigating a marketplace that moves faster than ever, but is still powered by human trust.

The forces shaping our clients' expectations are immense: climate urgency, data privacy, generational turnover, the social contract of remote work. Amid all that change, one truth remains: people still hire people. They don't hire and value relationships with avatars, chatbots, or dashboards. They hire belief and trust.

The invaluable skills that built your reputation—listening, clarifying, and translating—are the same ones that will defend it. What changes is scale. AI lets you listen to more signals, clarify faster, translate more dynamically. Used well, it amplifies your empathy. It does not replace it.

That's the opportunity this book explores: how to build smarter systems without losing the voice that built your business.

If *[re]wired* was about reconnecting the circuit after a financial shock, *[re]wired with AI* is about upgrading the circuit for a digital current. The goal isn't novelty for its own sake; it's longevity. To stay human, relevant, and creative in a world that won't wait.

We are not entering a "post-human" era of business development. We're entering a co-intelligent one. And that's good news because professionals already know how to collaborate. We just have to expand our idea of who's on the team.

How to read this playbook

If you've read any of my earlier work, you know I don't write books that are to be admired on the shelf. They're meant to live open on your desk and then annotated, dog-eared, and balanced on top of a Zoom microphone or travel bag. Think of this book less as reading material and more as an instrument you'll keep tuning as the tempo of your work changes.

Read it like a conversation, not a curriculum

Each chapter is designed to stand alone as a field note. You can read front-to-back or skip directly to the problem that's bothering you this quarter. Maybe it's a pursuit process that's grown stale, a team losing creative energy, or a marketing plan that sounds suspiciously like those of the last three years.

My tone will stay conversational and lightly urgent, though I'm not trying to lecture you. I'm trying to coach in real time, the way I do when a client calls and asks, *"We have a presentation tomorrow. What should we say first?"*

So, dip in and out. Start with a section title that matches your current frustration. Keep a pen or stylus nearby, because every few pages, I'll ask you to stop reading and do something. Experience travels faster through motion than through theory.

Watch for the sidebars. They're your built-in support crew

Throughout the book you'll see small markers:

- "Try This Now" One quick, usable exercise to translate ideas into behavior.
- "Human Moment" A reminder to reclaim empathy and presence when the workflow starts running on autopilot.
- "Toolbox" Curated prompts, templates, and technologies worth exploring. Each of these has been vetted against the rule, *"Would this actually help a busy professional tomorrow morning?"*

Use these as rest stops, not detours. They're there to convert ideas to action while your energy's still fresh.

Expect stories more than screenshots

You won't find loads of software tutorials here. There are already thousands of brilliant resources teaching you which buttons to press. What matters more is why to press them, and when a real human should step back in.

You'll see human stories made of small, recognizable moments from real firms. I've stayed far away from jargon and hero myths. You'll read about people who lost a pursuit, and then figured out why. About teams that turned a five-minute AI experiment into a companywide efficiency. About leaders who replaced panic with process.

These are modern parables for a modern industry. Read them quickly; discuss them slowly.

Use the frameworks as mix-and-match parts

The book's structure reflects the way real projects flow:

Part 1: *The Rewire: Mindset before mechanics*

Reset your mindset. Clear language precedes clear action.

Part 2: *Smarter Intelligence: From insight to expression*

Learn how to pair human judgment with artificial insight.

Part 3: *Winning the Moment: Proposals, pitches & performance*

Apply this intelligence to pursuits, proposals, and persuasion.

Part 4: *Leading teams through the change*

Build a culture that can sustain intelligent systems responsibly.

Part 5: *Sustaining the advantage*

Keep learning, sharing, and scaling without burnout.

Many parts end with reflection questions. Use them in team meetings or brown-bag sessions; they're built to spark discussion, not homework.

Make room for practice: Momentum arrives by doing

Every insight in the chapters that follow means little until you've watched it work in a live conversation. Try a small experiment first: rephrase one outreach email using the empathy framework, or use generative-AI to draft an outline before you brainstorm with your team. Then compare results.

That's how rewiring happens: one tiny voltage change, repeated deliberately, until your confidence becomes current.

Keep the goal in view

Ultimately, this is not a marketing book and not a technology manual. It's a performance guide for human design professionals living in and leveraging intelligent systems.

By the last page, I want you to feel two kinds of momentum:

1. The practical momentum that comes from reclaiming hours and focus through better tools.
2. The emotional momentum that comes from knowing your work still matters, maybe more than ever. Because machines can mimic knowledge, but only people transmit meaning.

If the first *[re]wired* helped you rebuild confidence after a financial shock, let this one help you rebuild that confidence after a digital one, and then go on to accomplish far more than you ever thought possible.

So read fast, annotate freely, share generously, and most of all keep the current moving forward.

PART 1

THE [RE]WIRE: MINDSET BEFORE MECHANICS

CHAPTER 1
The Rules Changed. People Didn't

Then vs Now

There was a time—not so long ago—that business felt like a slow, dignified waltz. You could build a reputation one handshake at a time. A client met you for lunch, saw a few of your drawings, liked the way you carried yourself, and by the following week the deal was yours. Submittals were printed, stapled, and couriered downtown. You knew exactly who opened the envelope: the same person who would later phone to say, *"Congratulations, you've got the project."*

Today, that same exchange happens through an automated portal, four log-in screens deep, with a confirmation email generated by a bot politely promising that "Your submission #17453 has been received." No one remembers who actually read it. Inboxes are flooded, decisions delayed, and our industry has responded with a frantic arms race of technology—campaign managers, analytics dashboards, digital-twin data, generative text engines promising salvation.

We've changed everything except the one thing that matters: the people. Because in every era, whether a proposal arrives by bicycle courier or through the cloud, the moment of decision still belongs to a human being with eyeballs at the other end.

That person has the same heartbeat of uncertainty, the same invisible checklist: *Do I trust them? Will they make me look good? Will this be safe for my career?*

The music of the dance has changed, but the steps haven't. What we called "relationship- building" in 2005 we now call "client experience." The labels have been updated; the instincts are ancient.

We've mistakenly believed that new tools create new rules, and that misunderstanding has kept many professionals exhausted yet strangely stuck. We chase the glitter of novelty while the substance of persuasion hasn't moved an inch since the first merchant laid out goods in a market stall and looked a buyer in the eye.

Ask any buyer today how they actually choose among qualified firms. They'll mention data, metrics, algorithms, ESG alignment, digital maturity ... and then, once the microphone is off, they'll admit: *"I went with the team I felt better about."*

That confession is not weakness. It's wiring. The limbic system still holds veto power over the spreadsheets.

So yes, we now brand ourselves with hashtags, measure engagement instead of eye contact, and fight to keep up with AI-powered competitors who can draft fifty variants of a proposal before breakfast. But the reader of each variant is still human. She still feels pressure, fatigue, ambition, and fear. And she will still, inevitably, award the work to the team that reduces those feelings, not the one that amplifies them.

The paradox of our new world is that while tools have given us more ways to speak, fewer of us are actually heard. Noise has multiplied faster than meaning. Clarity, empathy, and confidence cut through the marketplace as differentiators twenty years ago. And they cut through even faster now, precisely because so few competitors remember them.

So, this chapter begins there: at the intersection of permanent human wiring and the shifting abilities of technology. The rules of communication, procurement, and marketing have changed; the people operating within them have not. Recognize that, and every new platform becomes a translator, not a replacement, for the oldest rule in business: *A client buys what you're selling because they believe you understand them.*

The myth of the new rules

Every few years our industry seems to declare a marketing revolution. One decade it was the fax machine that would "eliminate distance." Then came the miracle of email newsletters, customer-relationship databases, search-engine optimization, social media, and now a chorus of AI prophets promising to write, design, and even feel for us. The proclamations always tout the same myth: *This changes everything.*

What actually changes, of course, are the *tactics*, in other words, the mechanics of how a message travels from you to the person holding the checkbook. What never changes is *why that person chooses you once the message arrives.*

Think back on the great disruptions in your own lifetime. Each one seemed to rewrite marketing physics for a moment. But when the dust settled, the winners had not mastered the newest platform; they had mastered *human motivation* inside that platform.

- The first firms to adopt email didn't simply send more correspondence. They wrote clearer notes and replied faster, signaling reliability.
- The first firms with websites didn't win because of animation. They won because clients could finally *see* proof of competence without having to ask for a brochure.
- Today's firms gaining ground with generative AI aren't those flooding LinkedIn with robotic posts. They're the ones using the tool to express their humanity more efficiently. They're freeing time to listen, not just broadcast.

Every era dresses up the same impulse in new clothes. The 1960s salesman, smoothing his tie before a pitch, was practicing the same art as the modern marketer checking tone and timing before hitting "send." Both are wrestling with one question: *How do I get someone to see what I see?*

The myth of the new rules tempts us to chase novelty instead of mastery. It whispers that if only we had the latest analytic stack or influencer strategy, the resistance would melt away. But there's a simple physics to persuasion: Attention flows toward relevance, trust, and clarity. Technology can accelerate or distort that flow, but it cannot replace it.

The modern proposal portal can collect fifty submittals at midnight. It cannot *care* which one feels right. That's still a human burden. The algorithms scour data to find signals; the nervous human being interprets what those signals will mean to their boss, their budget, their reputation. When the contract is awarded, the algorithm logs a transaction. But it's a person who feels the relief.

In countless workshops I've ask seasoned professionals what keeps them up the night before an interview. Their answers haven't changed in forty years:

- Will they like us?
- Will they trust that we can deliver?
- Will our price paint us as risky—too high or too low?

Notice the pattern: every worry circles back to emotion, not information. Fear of rejection. Desire for safety. Need for status. The same triad that sold wool in ancient markets still sells billion-dollar infrastructure today.

Technology has not altered human motives; it has merely made them *visible at scale.* Digital behavior data now graphs what was once private intuition: how often people reread an email before replying, where they hover on a web page, how quickly they abandon a video when it loses coherence. Underneath those graphs is the same pulse: a craving for reassurance, a search for meaning, a fear of wasting time or looking foolish.

When you accept that reality, technology becomes less intimidating and far more useful. It stops being an idol to appease and becomes a microscope for empathy. Every click, every dwell time, every AI transcript is a clue to the same constants we've always known: curiosity, insecurity, aspiration. Your job hasn't changed; the tools just hand you more evidence of why your work matters.

The only real "new rule" is that ignoring the old ones is more expensive than ever. The faster channels move, the faster misunderstandings spread. The leaders who prosper in this era are those who remember that the scariest headline in business development—*Our clients have changed*—is only half true. They haven't changed; they've simply gained more screens through which to reveal what they've always wanted: competence they can trust, people they enjoy working with, and the comfort of feeling safe in their choice.

Clients still buy confidence, safety, and fit

Strip away every procurement rule, every interface, every technical criterion, and every algorithmic ranking, and you'll still find one truth that no market disruption has dented: *clients buy peace of mind.* The labels might change. We can call it value, reliability, innovation, or ROI. But underneath each is the same three-part code that has guided human exchange since the first deal was struck across a table: *Confidence, Safety, and Fit.*

1. Confidence: Can they really do this?

Every project, no matter how routine, contains an invisible leap of faith. The buyer is betting their credibility and their internal political capital on you. Confidence is the antidote to risk. It's easy to assume clients buy competence, yet competence is merely the *entry ticket.* Confidence is what closes the sale. When everything about your message, your demeanor, and your process communicates "We've walked this path before and know where the cliffs are," anxiety drains out of the conversation.

Confidence means neither arrogance, nor perfection. It's conveyed through small moments: a proposal that anticipates a worry before it's voiced, a project manager who admits uncertainty but outlines how they'll resolve it by Tuesday, an email that answers the unspoken *what happens next?* question before the client can ask.

AI can summarize your experience more elegantly than ever, but it can't supply the single human quality that radiates confidence—*earned calm.* A calm that says, "We've seen this movie before, and everyone survived the ending."

2. Safety: Will this make me look smart or get me fired?

Early in my career, an older client confided after awarding us the project, *"I didn't hire you because you were the best. I hired you because you felt like you were the least likely to embarrass me."*

That line should be engraved above every proposal desk. In the calculus of selection, avoiding humiliation outweighs chasing brilliance. The world applauds novelty but buys security. Every buyer, particularly in the B2B or institutional space, filters risk through three lenses:

a) *Financial Safety:* Will this firm stay solvent and deliver on time?
b) *Political Safety:* Will my peers and superiors approve of this choice?
c) *Emotional Safety:* Will these people treat me with respect and keep me informed?

When your marketing, interviews, and digital presence signal stability without stiffness via your timely responses, consistent tone, and predictable follow-through, you become the safe harbor. Interestingly, technology amplifies this signal. In the age of AI, where so many messages feel synthetic, a voice that's warmly consistent now shines as authenticity itself.

The paradox is that automation has made humanity the scarce resource. Which means safety now sounds like *realness.* When you include a human-written aside, a small joke about traffic, a personalized comment on the client's recent post, these all communicate, *"This message wasn't generated; it was cared about."*

3. Fit: Do they get people like us?

Fit is chemistry translated into professionalism. It's cultural empathy and the sense that you belong in the client's world. When clients say, *"They just get us,"* what they're expressing is relief that they don't have to keep explaining.

Fit isn't achieved by mimicking jargon. It's built by curiosity and aligned values. In the AI era, data can tell you everything about your client's market but nothing about their culture. Algorithms analyze behavior; empathy interprets motivation.

That's why every tool you adopt should serve one purpose: learning faster what matters most to them. Keyword trend analysis, CRM scoring, or sentiment tracking mean nothing until you can answer the question, *"How do they define success, and can we celebrate it with them?"*

Fit is what transforms a transaction into a relationship. It's not about being identical to your client, it's about being complementary. The client builds confidence through your mastery, safety through your reliability, and comfort through the feeling that you *see* them. When those three align, the importance of price recedes, and loyalty becomes the natural outcome.

Every shiny innovation since the telegram has been marketed as a revolution in connection, and each time, the firms who succeed are those that make people *feel* connected. You can't automate confidence. You can't spreadsheet safety. You can't template fit. You can, however, use every modern tool from predictive analytics to generative language to *reveal* those qualities more vividly.

The portals and communication channels evolve, but the motives don't. Clients today may arrive through a QR code instead of a handshake, but what they're buying is the same: someone who lets them sleep through the night confident that what they care about is safe in capable hands.

Why this matters in the AI era

There's an old rule in architecture: *form follows function.* In marketing today, the corollary might be *tools follow temperament.*

The quality of what a new technology produces will never exceed the clarity and empathy of the person guiding it. Artificial intelligence is simply a mirror that amplifies whatever we bring to it. If our thinking is precise, the reflection will sharpen it. If our motives are muddied, the reflection will multiply the confusion.

The Amplifier, Not the Author

The early adopters of AI-based tools are discovering that they don't actually create; they intensify. Feed a half-formed marketing plan into them, and you'll get a dazzling, well-organized half-truth back in milliseconds. Feed them a crystal-clear message, rooted in understanding of what the client truly fears and desires, and the same engines will help you express that empathy with an eloquence and speed unimaginable a few years ago.

That's the hinge on which this new era turns. AI isn't another marketing channel. It's an amplifier of intent. It magnifies tone, logic, and structure. When your intent is to connect and reassure, it helps your meaning travel farther. When your intent is simply to impress, it turns vanity metrics into a wall of noise.

Speed Without Soul

Speed is intoxicating. The ability to produce a month's worth of content before lunch feels like progress until you notice the engagement graph: a brief spike, then silence. The missing ingredient isn't strategy or SEO alignment. It's you.

Clients are hypersensitive to tone now. They can tell when words were manufactured instead of meant. They don't mind if an algorithm assembled the draft; they mind when nobody took responsibility for its humanity.

Readers subconsciously scan your content for signs of life in the form of the unexpected idiom, the small confession, or the moment of humility that says, *"A person wrote this for another person."* Remove those, and even the most accurate message feels sterile. It's perfect, but unloved. In other words, while software can mimic competence, it can't replace your passion and commitment.

The Paradox of Data

For all the promise of predictive analytics, there's an irony lurking behind those dashboards. We now have *too much* evidence of behavior and not nearly enough understanding of motivation. You can measure views, clicks, scroll depth, and keyword resonance, yet still miss the heartbeat. Why? Because data tells us what people did. It will not explain *why* they cared long enough to do it.

Don't get me wrong, this isn't an anti-technology argument. It's a reminder that metrics are rearview mirrors. They show where attention traveled, not what guided the driver. The role of the human strategist is to interpret those signals through empathy: *What was happening in their world when they searched that phrase? What tension were they trying to resolve?*

AI can answer any question except the one it wasn't asked. It's only your uniquely human curiosity can frame that question in the first place.

Empathy as the new differentiator

During the technology shifts of the previous forty years, firms competed on efficiency—the ability to make more of the same thing faster. Now, efficiency is trivial, and the abundant product is content. But the scarce resource is *understanding*. Empathy, the capacity to read a person's unspoken context, has become the new premium.

As algorithms shoulder more of the heavy lifting, empathy will determine which firms still feel real. The proposers who consistently win will be the ones whose prompts are driven by humanity: *"Show me five ways to reassure a nervous facilities director,"* not *"Write five reasons to choose our firm."*

While AI can simulate credibility, it can't experience the feeling of relief. Your clients, on the other hand, thrive on that feeling.

So, the more we automate, the more our competitive edge shifts inward, from data mastery to emotional literacy. The question *"How well do you know your clients?"* no longer means *"How full is your CRM?"* It means *"Can you sense what matters to them before they articulate it?"* Technology can accelerate the delivery of empathy, it can't supply it.

The Human Circuit Breaker

Every firm now has access to the same algorithms, the same large language models, the same cloud horsepower. The firm that stands out will be the one that remembers to use the *off* switch. Human judgment is the circuit breaker that keeps automation from burning down trust.

Before you let any AI-generated word leave the building with your logo on it, pause and ask three human questions:

1. *Does this sound like us?*
2. *Does it sound like I understand them?*
3. *Does it make both of us more human?*

If the answer is no to any of the three, stop the automation loop and apply your human conscience.

The firms that navigate this transition successfully will not be the ones who mastered prompt engineering. They'll be the ones who mastered perspective and used machines to free their minds for the work that no machine can do: imagining, empathizing, deciding, caring.

We used to fear that AI would replace our creativity. The greater danger is that it replaces our empathy and our curiosity. Guard those, and every new technological ability becomes fuel for meaning rather than just more noise. Because, when the algorithms have saturated the world with content that sounds "right," the only voices still heard will be those that *feel* right. And that will always take a human to create.

Mini Case: When a firm chased data and lost its heart

A few years ago, as many companies outside the AE world (the most prominent of which was Google) were putting an increased emphasis on the value of data, a mid-sized interdisciplinary design firm in Colorado, made a very public commitment to "data-driven decision making." They called it their *Quantum Marketing Initiative*, complete with stylized dashboards that were shared with all employees on their intranet site.

To their credit, they weren't wrong about the world trending toward metrics; they simply misunderstood what the numbers were supposed to serve.

The Rise

The initiative began with promise. The firm retained a consultant with a strong tech background, subscribed to a half-dozen marketing platforms, and built out a gleaming new CRM that cross-referenced every client contact, proposal, and digital impression. They could now visualize their universe of prospects down to the minute. They knew who opened what, when, and for how long.

In the first six months, internal enthusiasm soared. Marketing reports were thick with heat maps and click-through rates. Staff meetings opened with charts of trending content. Management declared they were finally "learning what the market wants."

And yet, by mid-year, proposals were landing with a strange thud. Hit rates fell from 28 percent to 14. Pipeline velocity slowed. While the dashboards lit up, the phone didn't.

Team morale followed the metrics downward. Designers joked that the new CRM "knew more about our clients than we do—and likes them less."

The Fall

Six months into the initiative, and sensing that something wasn't right, the firm commissioned an outside audit. Instead of the hum of creativity and engagement the review found nothing but a hush as heads bent down to screens. Marketing staff were no longer telling stories, they were reporting observations. Proposals read like statistical reports filled with graphs of past performance, AI-generated paragraphs on efficiency, and not a single sentence that sounded like it had been written by a living, breathing human.

In chasing *data sophistication*, the firm had starved itself of emotion.

An honest conversation with one of their most loyal clients explained everything. When asked why the firm hadn't been considered for a recent hospital project, he spoke frankly:

"They had stopped talking about *patients* and instead were talking about *platforms.* I don't need a dashboard partner; I need designers who remember why hospitals exists." That line hit like a wrecking ball.

The Turn

To the firm's credit, they listened. A retreat was held in the foothills west of Boulder. Phones off. Laptops closed. They pinned every recent piece of marketing up on the wall and, one by one, asked, *Would this make any client feel seen?* When the answer kept coming up "probably not," they rolled up their sleeves and began a different experiment.

They rewrote everything—not from data, but from empathy. Outlines for each new proposal began with a statement of human stakes:

- *The mothers waiting in the emergency room at three a.m.*
- *The teacher unlocking the classroom on a winter morning.*
- *The technician whose shift ends at sunrise.*

Only after that did the team connect the statistics, charts, and predictive models.

They also instituted a new internal policy: every analytics report had to end with *one sentence of empathy*—a plain-English interpretation of what the data revealed about real people. For example, instead of *"30 percent higher engagement,"* the report read, *"Prospects lingered on the case study about flood recovery. It was obvious that they're anxious about their community."*

That shift reframed every meeting. Suddenly the dashboards were not gods to obey but microscopes to learn from.

The Result

Within two quarters, the firm's hit rate had climbed to 37 percent—its highest ever. But the leadership team noted something deeper. Clients were calling them again, and not just in response to submittals. Conversations returned to warmth and curiosity, not just cursory specifications. Staff talked about projects with pride instead of performance-metrics pressure.

Their technology hadn't changed. Their intent had.

When they stopped asking, *"What does the data say?"* and started asking, *"What are these numbers trying to tell us about the humans behind them?"* they turned the so-called sophistication into empathy, and the market felt the difference. Today, their *Quantum Marketing Initiative* is still alive, but the tagline on the office wall has been rewritten by hand: *"Data serves empathy. Empathy earns trust."*

It isn't as sleek as the dashboards, but it's infinitely more empathetic. And profitable.

Human Moment:
Re-centering empathy in a data-obsessed culture

When we're surrounded by charts, metrics, and automation, it's easy to forget what data is supposed to serve: understanding. Take one small pause today before writing or sending a report or proposal and ask yourself, *"Who's on the other side of this number?"*

Picture a single person affected by the project. Perhaps it's a facilities director, a teacher, a commuter, or a patient. Then reread your message through their eyes. If it still sounds mechanical, add one sentence that speaks directly to that person's world. That fifteen-second adjustment restores tone faster than any rewrite.

Practical Tool: The "3 Human Questions" checklist

At the end of every strategy meeting, marketing draft, or AI-generated output, there's a moment of choice. You can press *send* and hope the client sees the brilliance, or you can pause and ask the questions that really decide whether your message will land. These questions aren't about grammar or graphics; they're about reassurance. They're the *3 Human Questions* every client asks, whether silently or aloud, before committing.

1. Can I trust you? (Confidence)

This is the instant, almost instinctual, judgment your message triggers. Before a client reads line two, they're scanning for signs of steadiness: Is this firm credible? Are they consistent? Do they sound like people who deliver on promises?

To address it:

- *Show proof early.* Demonstrate reliability through short, specific examples—not grand claims.
- *Use calm language.* Replace "We're excited!" with "Here's what success will look like."
- *Be visually coherent.* Fonts, colors, and tone should sound like one voice, not a committee.
- *Acknowledge risk openly.* Stating what could go wrong and how you'll manage it builds more assurance than painting perfection.

If your draft feels breathless, defensive, or cluttered, the silent answer is *no*. They can't trust you yet.

2. Will you make my life easier? (Safety)

Every buyer is carrying invisible stress. A project isn't just dollars and deliverables; it's deadlines, stakeholders, politics, and personal reputation. Safety means, *"If I choose you, can I stop worrying?"* To reinforce safety:

- *Show understanding of context.* Signal that you grasp not only the technical scope but the organizational pressures.
- *Clarify communication rhythms.* "You'll get a progress note every Friday at noon." Simple and concrete.
- *Demonstrate competence without ego.* Modesty signals that you're in control, with nothing to prove. Bravado can read as desperate.
- *Sound stable.* Avoid hype words like "cutting-edge," "disruptive," and "game-changing." They sound risky, like you're trying too hard.

When your proposal feels like a relief to read, with smooth confident pacing and no surprises, you've succeeded.

3. Do you get people like me? (Fit)

A good fit begins the moment your reader stops judging and starts imagining a partnership. They see themselves shaking your hand at the ribbon-cutting or reviewing drawings over coffee. It's empathy made visible.

You demonstrate fit by:

- *Mirroring their language, not mimicking it.* Reflect their priorities such as schedule, quality, or community impact. But don't simply parrot jargon.
- *Referencing shared values.* If sustainability drives them, show how it genuinely drives you.
- *Personalizing details.* A brief nod to their city, staff, or recent initiative affirms that this communication could only have been written for *them.*
- *Including humanity.* A sentence that begins with "people who will use this facility …" immediately shifts perspective from specifications to the lived experience of real people.

When the fit is right, the reader stops comparing you to competitors and begins to compare you to their vision.

How to use the checklist

Before releasing any communication such as a proposal, an email, a presentation, or an AI draft, run through a 90-second gut check:

- *Does this sound trustworthy?*
- *Does this make the client's world lighter or heavier?*
- *Does it prove we understand their world?*

If you can't answer yes to all three, don't polish the layout. Rethink the message.

Used consistently, these questions transform marketing reviews from "Does it look good?" meetings into "Does it feel right?" conversations. They replace endless technical debate with a single purpose: to earn an enthusiastic *"yes!"* from a human.

Toolbox: Prompts and resources for human-centered writing

Here are a few easy tools to help make every communication just that bit more human.

A few prompts for everyday use

- *Clarify Tone*: "Rewrite this paragraph so it feels calm, direct, and human, not promotional."
- *Refocus Intent*: "What's the emotional need behind this client's request?"
- *Find Fit*: "List three ways to show this client we understand their world without repeating their words."

Some AI tools worth trying

- Claude or ChatGPT: for early-draft clarity checks.
- Hemingway App / Grammarly: to strip out jargon and over-complex phrasing.

Simple checklist card

Keep a sticky note by your monitor that reads:
Trust • Safety • Fit • Clarity • Tone

Every message that hits four out of five of those marks will already sound smarter—and kinder—than most of what lands in any inbox.

Try this now: Rewriting for human connection

This 5-minute exercise can help you re-tune your real message before technology amplifies it. Use this as your first maintenance check for empathy. This short exercise isn't theory, it's maintenance. Think of it as tightening the bolts on your communication machine before you let AI or automation drive it any faster.

Step 1: Pick one real message

Choose one piece of outbound communication you've written or approved in the last month:

- a proposal opening paragraph,
- a client-update email,
- a website "About Us" blurb,
- or even a LinkedIn post.

You don't need a masterpiece, just something you actually sent.

Step 2: Read it like a client

Forget who wrote it. Pretend you're the person on the receiving end, late on a Friday, scanning messages on a phone. Ask yourself:

1. *Do I trust these people?*
2. *Do they understand the world I'm in?*
3. *Do I feel calmer or more anxious after reading this?*

Mark the moments where the answer is *no* or *not sure.* Those are the fault lines where attention begins to slip and doubt creeps in.

Step 3: Apply the 3 Human Questions

Now, rewrite the piece guided only by the checklist from the previous section.

- *Trust:* Replace generic claims with verifiable proof or lived experience.
  Before: "We're the leading experts in structural systems."
  After: "Over the past decade, we've delivered 140 completed structural projects with less than 1% RFIs from contractors."

- *Safety:* Remove friction, uncertainty, and jargon that makes readers work too hard.
  Before: "We anticipate interface coordination challenges between MEP and envelope subs."
  After: "We'll host short weekly coordination calls, so field issues don't surprise anyone, especially you."

- **Fit:** Personalize the message for the real humans involved.
  Before: "Our firm prides itself on community focus."
  After: "Your downtown revitalization project shows exactly why we love community-scale work: buildings that bring the neighbors back out into the sunlight."

Don't worry about style at first; aim for sincerity. Correct English can always be edited later but a sincere sentence can't be faked by any algorithm.

Step 4: Read it aloud

When you hear your own voice delivering the message, the wrong notes are instantly obvious. If you cringe, it's not a punctuation problem, it's a tone problem. Adjust until it sounds like something you'd say sitting across a table from a trusted client, with a coffee or a beer between you.

Step 5: Share and compare

Print both the original and the rewritten versions and ask a colleague which they'd rather receive if they were the client. Ninety-nine percent of the time, they'll choose the second. The other one percent? They'll be secretly embarrassed that the answer was so easy.

The Lesson

Every technology revolution since the printing press has tempted professionals to trade *human clarity* for *mechanical efficiency.* That bargain always ends the same way: the clever tools remain, and the firms that remember humanity outperform those that forgot it.

You don't need to banish AI, automation, or analytics. You simply need to make sure that, underneath every campaign and every line of generated text, the message still whispers the three promises clients long to hear:

You can trust us.

You'll be safe with us.

You belong with us.

That's it. That's the algorithm that never changes.

Chapter reflection: The rules changed. People didn't.

Take a breath before you rush ahead. The purpose of this first chapter wasn't to teach new tricks. It was to *reset your instincts* before the technology fireworks begin. Everything that follows in this book builds on these constants.

1. Technology raises the volume, not the truth

Every new platform, from the printing press to AI, makes it easier to speak louder and faster. Only human intention decides whether that loudness becomes clarity or clutter. Before you integrate any new tool, ask: *Will this make my message more human or just louder and more frequent?*

2. Connection has always been emotional work

Markets evolve, procurement rules mutate, and project delivery methods morph. But trust, relief, and belonging still do the heavy lifting. The moment you forget that, your messaging slips into noise, no matter how sophisticated the metrics behind it.

3. Empathy is now a measurable advantage

Algorithms tally outcomes. People remember how you made them feel *while* achieving them. The best marketers in the AI era won't be those who use the most technology; they'll be those whose technology delivers the most empathy per click.

4. The Three Human Questions run the show

Every open rate, every short list, every signed contract is an unconscious *yes* to:

- *"Can I trust you?"*
- *"Will you make my life easier?"*
- *"Do you get people like me?"*

Put those three at the top of every creative brief, and you'll spend less time persuading and more time partnering.

5. Measurement without meaning is just math

Data is essential, but it's not insight until you attach a human story to it. Instead of, *"Our proposal views are up 25 percent,"* try, *"More people lingered over the testimonial about flood recovery. It seems that that's the issue keeping them up at night."* That simple reframing turns analytics into empathy.

6. Your future tools depend on your present character

AI is an amplifier. It will exaggerate your best strengths and your smallest indifferences. If you aren't grounded in curiosity, humility, and genuine care for the person across the table, the technology will only make that absence more obvious.

7. The timeless equation

Every project you'll ever encounter will always be one human asking another, *"Can I trust you with something that matters?"* That question will never disappear and marketing will always remain a practice of establishing faith under pressure.

Remember: As you go forward in this book, you'll be exploring a new set of tools. But your compass must never change. While the coming chapters will introduce automation, analytics, and AI workflows, each one will circle back to the fundamental constant:

Marketing succeeds when it makes humans feel safe, seen, and sure.

CHAPTER 2

What AI really means for AE marketers and sellers

An anecdote: Surprised by a machine

Maya had a 45-minute window before a call with a prospective client in Austin—a health-care system wanting to explore planning work for new facilities. She'd skimmed her firm's brochure, glanced at the hospital's website, and realized she didn't know much about the client's pain points.

Curious, and pressed for time, she opened ChatGPT. Her prompt was simple:

"I'm an architect preparing for a 30-minute intro call with a regional hospital client. What five questions might help me understand their top facilities challenges?"

Ten seconds later, a list appeared on her screen: *Ask about staffing shortages, deferred maintenance, infection-control upgrades, patient-flow bottlenecks, and energy resilience.*

Nothing magical so far, as these were all known issues. But the phrasing caught her eye. Each question was written conversationally, patient-first, not architect-first. And it was that subtle shift to the human cadence she hadn't expected. And that changed how Maya approached the meeting.

A small discovery. A large realization

The call went better than most. She talked less, listened more, and ended with the client saying, *"I feel like you actually understand the pressures we're under."*

Later she admitted to a colleague, half-embarrassed, *"I didn't use the machine to write* for *me. It just reminded me how to sound more like a human."*

That moment captures where we stand today in the swirl of AI headlines. Not a future of robot strategists replacing marketers, but one of digital mirrors helping people rediscover their own effectiveness. It's humbling, sometimes unsettling, and surprisingly practical.

Why start here?

This small story contains the paradox the whole industry is wrestling with: The more the tools evolve, the more they nudge us back toward the human basics we'd been too busy to practice.

AI overawes us with its speed and scale, yet its first meaningful impact for most professionals isn't revolution, it's reflection. It shows what clear, empathetic communication looks like when stripped of hesitation and ego. And that's the right emotional entry point.

Before we can use AI wisely, we have to *see it accurately*. Not as magic, not as menace, but as a new class of helper that imitates language, not thought.

Where we're going next

Now we're going to slow down and determine what, exactly, this "helper" is doing when it seems to know so much. We'll define AI in the plain language of pattern recognition and language generation and set a calm, factual baseline before discussing what's possible, what's risky, and where people remain in charge. Every confident user of new technology begins with understanding, not awe.

***

Defining AI in plain language: Pattern and prediction

When people talk about *artificial intelligence*, most imagine a kind of digital brain—something that can think, reason, and maybe, someday, replace us. That image is wrong, and dangerously distracting.

Real AI, the kind you're already using in everyday tools, doesn't *think* at all. It *recognizes patterns* and *predicts what usually comes next.*

That's it. And that's enough to be powerful.

An honest definition

In plain English: *AI is a system that finds patterns in past data and uses those patterns to predict or generate likely outcomes in the future.*

When you use ChatGPT, Midjourney, or an AI search assistant, you're not talking to a mind. You're talking to a *statistical mirror* trained on enormous volumes of text or images, finely tuned to continue a pattern that feels human. It works by probability, not understanding.

If you type *"Our client is building a new downtown library…,"* the model has read millions of similar sentences and computes what's most likely to follow: *"how could we design it for community engagement?"* It's not being insightful, it's being *statistically fluent.*

Most of today's "generative AI" tools run on this simple principle, though their training data and interface differences make some feel more creative than others.

From pattern recognition to language generation

Two steps power almost everything you see:

1. **Pattern recognition**

The model scans examples. It learns relationships and how certain words (or pixels, notes, data points) relate to others. It figures out rhythm, tone, even emotional temperature, simply by counting and weighting co-occurrences.

2. **Language (or image or data) generation**

When prompted, it predicts the next most probable word, pixel, or outcome, one unit at a time, using those learned weights. That's why the output reads smoothly, even creatively. But it's nothing more than a plausible continuation of patterns. It's like the autocomplete that engages when you write an email or text, but evolved to genius level.

Why the pattern lens matters

If you remember this one concept—pattern and prediction—you'll make far saner, safer decisions about AI adoption.

- **It will stop you from expecting too much.**
  If you know it's only predicting and not reasoning, you'll double-check before you copy-paste.
- **It will stop you from fearing too much.**
  If it's only reproducing patterns, you realize that your judgment surrounding the context, taste, and empathy remains the decisive advantage.

Marketers and sellers live precisely in the gaps that AI can't reach as they modify choices with emotion, timing, and trust.

A useful analogy

Think of a seasoned jazz musician. After years of hearing progressions, they instinctively know which chord likely follows another.

AI is an instrument that's memorized every chord change in history, but it's never actually *played a gig*. It can riff convincingly, but it doesn't know whether the audience looks engaged or bored.

That feedback loop of human response is your domain, not the machine's.

What this means for your work

When you use AI in your marketing or business development, you're not summoning intelligence; you're *collaborating with probability*. The better you define the boundaries and intention and the more context and restraint you supply, the more useful that collaboration becomes.

The danger lies not in the tool's imagination, but in our willingness to accept its confidence as competence. When you understand AI at this human-scale, that it is finding patterns, not thinking, you replace both hype and dread with a manageable word: *precision.*

Now that you see what's *really* under the hood, we can move into the practical continuum: from the minor assistants you're already using without noticing, to the sophisticated systems that are capable of reshaping proposal work, client targeting, and idea exploration.

The continuum of tools: From email drafts to proposal intelligence suites

Most people talk about AI as if it's a single thing: a monolith of code and mystery. In reality, AI shows up on a continuum, from the invisible helpers already inside your inbox to enterprise-scale systems predicting billion-dollar trends. Seeing this continuum clearly helps you separate hype from progress.

It also lowers your blood pressure. Because when you realize you've been living with AI for years, the unknown stops feeling so mysterious.

1. Ambient AI: The helpers you hardly notice

If you use Gmail, Outlook, or Shopify, you're already swimming in soft AI. Every time your email client suggests a reply *(Thanks, sounds good!)* that's machine learning as it detects intent and predicts next words. Every time your phone camera refocuses a face, or your calendar auto-schedules based on time zones, that's pattern recognition at work.

These are not "giant language models." They're micro-learning systems optimized for one purpose: to save people a few seconds or clicks. They're friction reducers, not decision-makers.

For busy marketers and sellers, these small AIs quietly reclaim minutes that add up to hours. They don't write strategy, but they create room to think strategically.

2. Generative assistants: The new creative co-workers

The next level up is what Maya used in our opening story: tools like ChatGPT, Claude, Gemini, or Copilot. They're powered by large language models (LLMs) that generate plausible text or images based on your prompts. Use cases are multiplying daily:

- Drafting email and social copy.
- Summarizing research or meeting notes.
- Translating technical project details into client-friendly language.
- Brainstorming story lines or proposal themes.

These systems are generalists: incredibly helpful, occasionally misguided, and entirely dependent on your direction. They feel magical the first week, annoying on the second, mechanical by the third, and invaluable once you learn to guide them.

Think of them as a *new intern who writes quickly but doesn't fully understand your business.* Useful? Absolutely. Trustworthy without supervision? Not yet.

3. Specialized engines: AI tailored for your workflows

A step further along the continuum are *domain-specific systems*: AI that is integrated into design, CRM, and marketing-automation platforms.

Examples include:

- *CRM predictive scoring:* Highlights which clients are most likely to re-engage, based on historical behavior.
- *Proposal-intelligence suites:* Compare language across past wins and losses to flag themes correlated with success.
- *Marketing-performance optimizers:* A/B-test visuals or headlines and adjust spending automatically. Sellers who use Amazon Ads are using these bots more and more.

- *Design generators:* In early conceptualizing, tools that render quick spatial layouts or visual mood studies.

These AIs don't try to mimic conversation. They crunch patterns in your own firm's data to find edges: what wins, what triggers engagement, what patterns predict risk. And because they use your history, *they mirror your firm's habits*, both good and bad. If your data is scattered, the insight will be too.

4. Strategic engines: The emerging frontier

Finally, beyond what most firms deploy today are *strategic-level AIs*. These are systems aiming to support pricing, forecasting, scenario planning, even ethical risk analysis. Right now, they live mainly in pilot programs or enterprise labs, but the direction is clear: tools that *advise*, and not just *assist*. They won't replace leadership, but they'll test leadership assumptions against a broader pattern library.

Smart, progressive firms are already experimenting in low-stakes areas. They are earning on marketing data before letting the models anywhere near client deliverables.

The promise is consistent insight without consistent burnout. The risk, however, is treating pattern as prophecy.

Why the continuum matters

Seeing AI as a range instead of a revolution lets you retain agency. You can decide how far along that line you want to travel this quarter, this year, or this strategic cycle.

Step one*:* Acknowledge the AI that's already saving seconds.

Step two*:* Intentionally explore generative tools with guardrails.

Step three*:* Pilot specialized engines tied to measurable outcomes.

Step four*:* Watch the frontier, but adopt new tools only when the purpose is clear.

This reframing puts technology back in its rightful place: *a servant to clarity, not a source of chaos.*

Now that we've located ourselves on the map, the next question is practical: *What do these tools actually do for your firm?*

In the next section, we'll cut through the abstractions and look at the *real business benefits*, speed, pattern insight, and creative lift. With these, the excitement around AI turns into measurable value, not another line item of good intentions.

***

The real business benefits: Speed, pattern insight, and creative lift

It's easy to dismiss AI as the latest shiny object. But beneath the buzzwords, there's a set of real, measurable gains already unfolding inside firms that have started to use these systems intentionally. They're not science fiction. They're workflow improvements that compound into strategic advantage.

When we separate principle from panic, three primary benefits stand out that can bring huge value to your marketing and sales process: *speed, pattern insight, and creative lift.*

1. Speed with discernment

Every marketer and BD professional knows how much time is lost to the "blank page." First drafts of proposal sections, sensitive emails, or thought-leadership posts can consume an hour just to gain traction. AI can compress that slog into minutes, producing a credible starting point without the usual inertia.

But the competitive advantage isn't just faster writing. It's *faster iteration.* It's the ability to generate, test, and refine multiple angles before your first coffee break.

Example: A pursuit manager feeding a project description into a language model can surface five positioning statements in seconds, then quickly test the most resonant one internally.

That acceleration means the time that people spend can re-center on review, nuance, and relationship. Which are the places where the firm's real value lives.

Used wisely, speed doesn't reduce quality. It redistributes energy toward higher-order thinking.

2. Pattern insight: Turning data into direction

Traditional analytics tell you *what happened.* AI, trained on the right data, hints at *why it happened* and *what might happen next.* In marketing and business development, that shift is nothing short of revolutionary.

Pattern insight allows firms to:

- Detect which proposal content correlates with shortlist rates.
- Spot seasonal or sector-specific engagement spikes.
- Identify under-nurtured clients who behave like your best ones.
- Recognize which content themes lead to inbound queries within 30 days.

While most of this has been theoretically possible through manual reporting, marketers and executives are too busy to slog through the data. AI reads faster, correlates deeper, and notices those valuable insights that fatigue usually makes invisible.

When machines find patterns, people regain the bandwidth to interpret them and decide which insights are meaningful, ethical, or even desirable to act on. That's how data becomes wisdom.

3. Creative lift: Breaking routine thinking

For all its cold logic, one of AI's most valuable roles is as a *creative disruptor.* While it doesn't "invent," it recombines ideas at a scale no brain could manage in isolation. When used intentionally, that randomness is fuel for originality.

- Staring down a stale campaign? Ask an AI to restate your message for "an audience of skeptical nurses" or "a plant manager worried about risk."

- Looking for a better hook? Have it list metaphors for "trust" or "momentum" and watch what shakes loose.
- Need to humanize a technical paragraph? Ask the tool to "translate for a community newsletter" and study the tone it produces. (I'll show you a real example of this later in this chapter.)

While it's rare that you'll use the output verbatim, the *distance* from your first draft to the machine's suggestion often sparks a leap your brain wouldn't have taken alone. That's creative lift: using the machine's acceleration to generate perspective you can then refine.

Cumulative effect = Strategic agility

None of these benefits alone rewrite a business model. But together they create what high-performing firms thrive on: *momentum without burnout.*

The rhythm changes. Less stalling, more iteration. Less guesswork, more grounded risk-taking. That momentum compounds into a real cultural advantage with a team that experiments, observes, and adjusts faster than competitors who are frozen by uncertainty.

A firm that knows how to harness AI responsibly isn't just more productive, it's *more adaptive.*

A simple rule of thumb

If you're trying to decide whether an AI application belongs in your workflow, test it against this question: *"Does it make us both faster <u>and</u> wiser?"*

If the answer is yes, explore further. If it only makes you faster, pause. Because speed without learning is how firms burn bright, then burn out.

Of course, every tool that amplifies strength can amplify error. The same systems that accelerate creativity can also flood your pipeline with beautifully written nonsense. Or worse, confidential data leaks and invisible bias. Which is why we need guardrails.

Next, we'll confront the real limitations and risks: What AI does poorly, what it misunderstands, and how to protect both your firm's integrity and your clients' trust while experimenting boldly.

Limitations: Hallucination, bias, and confidentiality risk

Every new technology begins with the promise of liberation. Then reality shows up. The truth is, while AI can accelerate and inspire, it's still fallible in ways that matter to you, your clients, your reputation, and your ethical accountability.

Understanding the limitations doesn't reduce opportunity. It *grounds* it, replacing naïve enthusiasm with confident discernment.

Let's look at the three core risks every AE professional should understand before using AI in business development or marketing.

1. Hallucination: When confidence outruns truth

"Hallucination" is AI's technical term for making things up. It happens when a model fills gaps in its training data by inventing plausible, but untrue, information.

Ask a model about a hospital project that never existed, and it may fabricate one, complete with location and a design award. Worse, it will do so with absolute composure, because language models have no internal truth-meter; they optimize for *fluency*, not *fact*.

Why this matters

In marketing and sales, persuasive tone is everything, which makes hallucination especially risky. A line of copy or a client reference that *sounds* credible isn't always *verifiable*, and publishing unverified content under your brand can shred trust faster than any typo.

How to protect yourself

- Never rely on AI-generated facts without human verification.
- Treat every "citation" or project reference as *potential fiction until verified.*
- If you can't confirm a detail, re-frame or write contextually ("Hospitals today are prioritizing..." instead of "Hospital X achieved...").

Machines don't mean to lie. They simply don't know what a lie is.

2. Bias: When patterns reflect prejudices

AI learns from mountains of past data, and that data carries society's fingerprints. Whatever cultural, racial, gender, geographic, or other patterns exist show up in the model's predictions. That's how hiring tools historically down-ranked women, or image generators defaulted to certain skin tones for certain professions.

In marketing language, bias can sneak in more subtly:

- Defaulting technical expertise to "he."
- Treating Western design norms as universal.
- Framing "cost efficiency" as more valuable than community impact, simply because the training data did.

Why this matters

A biased pattern, multiplied through automated content, can accidentally reinforce the very inequities firms claim to challenge. Bias isn't a public-relations problem; it's a values problem, and it originates in data, not intent.

How to counter it

- Ask tools to *diversify outputs* ("Offer three perspectives from different audience viewpoints").
- Build internal review checklists for tone, representation, and language.

- Use bias as a *teaching moment.* When you see it, you've found where your firm's ethical oversight should focus.

Bias isn't eliminated by better technology. It's balanced by better stewardship.

3. Confidentiality risk: When convenience leaks

AI systems learn from whatever they're fed. If that system belongs to a public platform, your inputs can be stored, logged, or used for further training, even if they're anonymized. This is one of the most underestimated business risks of today: *the accidental disclosure of sensitive client or project information through casual prompting.*

Marketing teams often paste snippets of proposals, client names, or pricing structures into chat interfaces to "clean up copy." That's how trade secrets, NDAs, and client confidences quietly enter someone else's server.

How to guard against it

- Never input confidential or identifying data into public models.
- Use private or enterprise-licensed versions with explicit non-training assurances.
- Sanitize project details ("Client A, a regional healthcare firm…") before seeking AI assistance.
- Train staff that *hygiene* in your *prompts* is now part of professional ethics.

AI is a power tool, not a private notebook. Handle it accordingly.

The deeper risk: Overconfidence

These three issues, hallucination, bias, and confidentiality, have a common root: human overconfidence in what looks polished. Because AI outputs *feel* articulate, we forget how brittle the reasoning behind them can be. That's why the smartest professionals now pair every test with a mantra: *"Check the output like it came from an intern who is smart, fast, but unlicensed."*

Trust its effort, but verify its evidence.

Risk is the price of capability

The point isn't to retreat from AI. It's to approach it with the same rigor you'd demand of any consultant, subcontractor, or junior staffer handling sensitive material. Risk, clearly understood, becomes *manageable*. What remains unspoken or unexamined can quickly become corrosive.

That's the leadership opportunity for every marketer and seller in this new terrain. You must be the calm translator between innovation and integrity.

So, if AI can accelerate us and also expose us, where's the balanced frontier? How do humans remain the decision center while letting machines lift routine work?

That's what we'll explore in our next section, showing how professionals can keep cognitive responsibility and creative authenticity while welcoming intelligent assistance.

***

The "Augmented You" model: Keeping the human driver in the loop

When new technology shows up, professionals tend to lurch toward one of two extremes: panic that they'll be replaced, or bravado that they don't need to change. Neither approach survives contact with reality.

The middle ground, the one where real growth happens, is the ***Augmented You*** model in which you accept that AI can assist, accelerate, and expand your capability, while your brain remains the command center. This isn't science fiction. It's the next evolution of professionalism.

The core idea: AI doesn't replace expertise, it extends its reach.

In this model, the people remain the *authors of intent* as they decide what questions matter, what patterns are meaningful, and what tone aligns with your brand. The machine becomes an **instrument**, turning that intent into faster drafts, broader insights, and creative variations.

Think of it like this: your calculator didn't replace math. It freed you to think about better problems. AI plays that role for language, strategy, and pattern recognition.

Three layers of augmentation

1. Cognitive extension: Think wider, not harder

AI can recognize and flag connections across vast datasets that no human could read in a lifetime. That doesn't make it wise, but it makes it *useful* as a thinking companion.

- Ask it to list assumptions you might be missing before a client call.
- Have it summarize differing viewpoints on a trend so you see the chessboard more fully.
- Use it to simulate how different audiences might react to a message.

You're still doing the judgment work. AI just widens the field of consideration so your conclusions are better informed.

Benefit: less tunnel vision, more situational awareness.

2. Creative extension: Explore without exhaustion

Many design professionals stop imagining too soon because creativity takes energy. AI can produce endless permutations of taglines, story angles, and client narratives, letting you explore breadth before you commit to depth.

The *Augmented You* uses that abundance not for substitution but for stimulation. You *borrow sparks*, not sentences. Using prompts such as: *"Generate three distinct tones for this message: confident, empathetic, and visionary"* helps you test emotional registers quickly, preventing your fatigue from narrowing your imagination.

Benefit: more creative range, delivered at the speed of curiosity.

3. **Operational extension: Automate without abdicating**

Many administrative drains such as scheduling emails, summarizing meetings, or transcribing brainstorm notes, are perfect for AI support. The key difference between automation and abdication is awareness.

Automation saves clicks while abdication ignores consequence.

The *Augmented You* knows which tasks are safe to delegate and which require judgment or contextual nuance. That awareness is the line between efficiency and error.

Benefit: reclaimed hours for strategic and relational work, without diluting accountability.

Guardrails for staying in control

1. **Always start with intention.**
   If you can't name the purpose of using AI in a task, you're not ready to use it.
2. **Insist on visibility.**
   You must be able to trace how an output was generated or at least know which system produced it.
3. **Validate before publishing.**
   Every output deserves the same scrutiny as any work authored under your logo.
4. **Train your team in hygienic prompting.**
   Treat sensitive data like client conversations. They're need-to-know only and never pasted carelessly into tools.
5. **Stay curious, not credulous.**
   Curiosity expands skill; credulity contracts it. Use the tool as a collaborator, not an oracle.

These habits hardwire your professionalism into your newly augmented practice.

A mindset shift more than a skill shift

The biggest leap isn't technical, it's psychological. Professionals must get comfortable with partial automation without surrendering ownership. That means unlearning an old lesson in which we were all indoctrinated: that doing something manually proves expertise.

In the AI era, expertise lies in knowing what to automate, what to interpret, and what to protect. Mastering that balance turns you from technician to orchestrator. You become a conductor of both human and machine instruments.

Two quick illustrations

First, let's visit Matt, a talented Civil Engineer who's as left-brained as can be. He was asked to write a project description for a treatment plant upgrade that he'd recently completed. Matt loves to do engineering, but cringes every time he's asked to write something. Instead of writing the description, he checked his project notes and dictated the following voice memo:

The Aerobic Digester Replacement Project incorporated a new covered reinforced concrete and masonry aerobic digestion facility, a new 2 meter gravity belt thickener, relocation of the existing 2 meter belt press, feed pump, grinder and appurtenances from the existing dewatering building to the new aerobic digester facility, addition of a new spare press feed pump, space for a future second 2 meter press, new dry and liquid polymer systems for the belt thickener and belt presses, shaftless screw conveyors to deliver sludge cake from the press to either of two truck or dumpster loading bays, air blowers, electrical room, storage room, yard piping, demolition of the existing aerobic digester.

Matt emailed the voice memo to the Marketing Coordinator and went back to work on his current project, which was definitely the highest and best use of his time and talents. The Marketing Coordinator took the (almost unintelligible) description and gave both Claude Sonnet and ChatGPT this prompt: *Rewrite this engineering project description for use in a community newsletter.* In less than ten seconds, the AI bots had returned these rewrites:

Claude Sonnet 4.5

Wastewater Treatment Plant Upgrades Complete

The town has completed major improvements to its wastewater treatment facility with the Aerobic Digester Replacement Project. This project replaced an aging digester with a modern covered building where beneficial bacteria break down solid waste materials.

The new facility consolidates equipment that was previously spread across multiple buildings, making operations more efficient. Key upgrades include new machinery for removing water from waste solids, updated pumps and grinding equipment, and modern conveyor systems that transport processed materials to waiting trucks for disposal. The facility also includes room for additional equipment as the community grows.

These improvements will help the town process wastewater more reliably and efficiently for years to come. The old digester has been removed and the site restored.

ChatGPT

Aerobic Digester Replacement Project

Big improvements have been completed at our wastewater treatment facility! The new aerobic digester building features stronger construction, better equipment, and more efficient systems to handle wastewater safely and reliably.

Crews have installed new pumps, presses, and conveyors that move treated material more cleanly and efficiently. The project also makes room for future upgrades, keeping our system ready for years to come.

Once everything is up and running, the old digester will be removed—making this a major step forward in keeping our community's infrastructure modern and dependable.

The Marketing Coordinator kept both versions, knowing that they were each suitable to an end use they had in mind.

Second, let's return briefly to Maya, our architect from this chapter's opening story. Three months later she's using AI every week, but not to write for her. She prompts it to *challenge* her thinking before major pursuits:

- *"List five objections a skeptical client might raise to our proposal."*
- *"Rewrite this paragraph at a fifth-grade reading level to test clarity."*

Each time, the system acts like a mirror, not a megaphone. She sees blind spots sooner, polishes messages faster, and walks into meetings better prepared.

These examples are the *Augmented You* in practice, a professional amplified by insight, not replaced by software.

To make this real, every professional needs a low-stakes way to experiment safely, without breaching data, confusing clients, or overreaching capability. That's the "lab" mindset we'll cover next: how to build a small, private sandbox to explore AI responsibly before systematizing its use.

***

Try this now: Build your own AI test lab

At some point, the only way to quiet the noise around AI is to experience it under your own fingertips. Not with high-stakes, real-world actions that risk a client deliverable, but a *sandbox* where curiosity is safe, data is clean, and learning feels playful again. So, I invite you to set up your own low-pressure test lab in which understanding can replace apprehension.

Step 1: Keep it personal and private

Start small and stay clean. You don't need enterprise tools or licenses at this stage. Just choose a single, reputable AI assistant you trust (ChatGPT, Claude, Copilot, whichever your firm approves). You can use the free version of any of these for now.

Then draw a clear boundary: *no client names, no pricing, no proprietary data.* Instead, use generic placeholders like *"Client A, a regional developer,"* because this is about *practice*, not production. You're not trying to prove anything yet. You're simply developing fluency with the tool.

Step 2: Pick three tiny experiments

Think practical, not grand. Choose three everyday marketing or BD tasks you already do weekly and see how AI handles them. For instance:

- Ask it to rewrite a dull paragraph from an old proposal into three alternate tones: informative, conversational, and visionary.
- Have it summarize an article or market report and explain what aspects of the content would be useful to your clients.
- Prompt it to brainstorm headline ideas for an upcoming email or social media post.

The goal isn't brilliant copy, it's *curiosity with feedback*. Notice what it gets right, what it misses, and more important, what helps you think faster or clearer. That's the exercise: using the machine to practice discernment.

Step 3: Practice prompting like you talk

Here's the secret I've learned: the best prompts sound like real conversation. You'll get better results when you speak to AI as if you're chatting with a smart, curious colleague. Casual, complete, and contextual. Try this shift in tone:

Flat Prompt	**Conversational Prompt**
"Write marketing copy for my firm."	*"Imagine you're helping me write a short paragraph to describe our architecture firm to a new client. We care about community impact more than awards, so keep it warm and grounded."*

Notice how the second one brings your *intention* into the request. It tells the model not just *what* to write, but *why* it matters.

These three small tips can make a big difference:

- *Be direct and human:* "Hey, here's what I'm trying to do..." works better than bullet-point lists.
- *Give context over commands:* "We're drafting this for a healthcare client, so clarity and trust matter most."
- *Request tone and role:* "Respond as if you're my marketing partner, not a robot consultant."

The conversation you start determines the quality of the help you receive.

Step 4: Capture more than outputs

As you test, jot down short reflections. Not what the tool produced, but *how it changed your thinking.* Did it help you start faster? Spot a better angle? Clarify something you didn't realize you were fuzzy about?

Don't just save what the tool generates. Save your reactions. Note when it surprised you, disappointed you, made something clearer, sparked an idea. Those behaviors are the real takeaways. They're the growing sense of what a healthy collaboration with AI feels like. In time, those notes become *your internal manual* for how you work best with AI.

Step 5: Invite one colleague

Learning alone is useful. Learning *with* someone is both safer and more fun.

So, find a test buddy. Swap one safe prompt a week and talk about what changed when you each asked the same question in different ways. The contrast in how other people use the same tool and shape their prompts will tune your own instincts faster than any formal training. It also turns learning into shared discovery rather than solitary trial and error.

Remember: this isn't about perfecting prompts; it's about building *reflexes of curiosity and accountability.*

Step 6: Set your own ethical fence

As your lab experiments expand, pause to define your personal guardrails, the "no-go zones" that keep you trustworthy (e.g. "I don't input pricing," "I don't co-author client emails," "I fact-check all statistics manually"). Ask yourself:

- *What will I never enter in a chat?*
- *When does human review become mandatory?*
- *How will I flag anything that feels off-message or off-values?*

Boundaries let your explorations turn into confidence. They make you the kind of professional your firm and your clients can trust to innovate *without gambling their data or dignity.*

The payoff

After a few weeks of light experimentation, you'll feel something subtly shift. AI will stop being *a mystery outside* your control and start becoming *a valuable skill in* your toolbox. You'll begin to hear its rhythm, recognize its limits, and spot that pivot point where *you* take over again.

That's fluency. It's the sign you're not chasing the trend anymore. You're shaping it. And that curious, experimental, and ethical mindset is how early adopters become future leaders.

Then, when colleagues inevitably ask, *"Aren't you worried this stuff will replace us?"* you'll be able to assure them: *"I was. Then I learned how to make it work for me. Now I'm using it mostly as a conversation partner. It helps me think faster and see wider. Then I make the calls."*

That's the *Augmented You* in real life.

***

Now that we've demystified what AI is, what it can (and can't) do, and how to work with it responsibly, we can tackle the emotional current running beneath all of this: the fear, excitement, and confusion that still colors every conversation about change.

That's where we go next. We're going to look at the fear, hype, and opportunity that surrounds AI. We're going to learn to spot the narratives that keep teams paralyzed, and how to replace them with stories that move them forward.

CHAPTER 3
Fear, Hype, and Opportunity

The temperature in the room

Mention AI in most business meetings today and you can feel the room tighten a little. People stop scrolling, fold their arms, look to see who will speak first. It isn't hostility. It's caution. Most of us have lived through enough technology cycles to know that enthusiasm and disruption usually arrive as a pair. The result is a mix of curiosity and doubt that you can almost measure in the silence before anyone answers.

Two common fears

When I ask teams what keeps them uneasy, nearly every concern traces back to two simple ideas.

Fear 1: I don't want to be replaced.

Behind that lies the old question of whether machines will take over the work people care about most. The evidence so far suggests something more complicated: *roles change faster than they disappear.* AI is already shifting what counts as valuable time, allowing less repetition and more synthesis. But the human elements of persuasion, empathy, and judgment remain the differentiators. Learning how to use new tools is still safer than hoping they go away.

Fear 2: I'll never catch up.

It's easy to assume everyone else knows what they're doing. The truth is, almost no one does. Every firm is experimenting.

Policies are evolving, comfort levels vary, and even the experts are testing boundaries as they go. Once people recognize that, the anxiety drops. Change becomes something shared, not something they're facing alone.

The role of emotion

AI touches more than process; it brushes against identity. Competent professionals don't like feeling unsure, and many worry that asking basic questions will make them look like they're behind the curve. That's not weakness, that's normal. Leaders who acknowledge that tension early make it easier for teams to move forward together.

What hype does

Public conversation still swings between extremes: full automation on one end, looming catastrophe on the other. Most of the real work sits in the middle with people adapting tools to their workflows and improving efficiency piece by piece.

Paying attention to that middle ground is how progress happens without panic. The trick is to stay skeptical of the mood that the headlines foster and stay focused on what actually works inside your own organization.

Reframing the moment

Instead of asking whether AI will replace us, the better question is, *"How it will change what we value and how we deliver it?"* Seen that way, the task ahead isn't defensive, it's professional: understand the tool, learn its limits, and decide where it belongs in your process.

If you've been around this profession for more than a minute, you've already experienced transitions like this before. The introductions of CAD, 3D Modeling, BIM, Lidar, CRM, and all the other disruptive technologies each caused its own earthquake. Whether or not you collapsed from the shaking depended on your mindset. That mindset—neither excitement nor fear—is what makes adoption go smoothly.

In retrospect, and separate from any emotion or sentimentality, the introduction and adoption of each of these technologies, has made the profession stronger, more capable, and better than before.

Next, we'll look at what this means for executives, managers, and project leads as we explore how to create an environment where people can test new ideas without risking quality or credibility.

***

Leading through uncertainty: Creating safety for curiosity

For most leadership teams, the hardest part of this transition isn't adopting new tools. It's managing the uncertainty that comes with them.

Engineers, architects, and marketers alike are used to clear deliverables and well-defined scope. AI unsettles that structure by introducing speed, ambiguity, and results that vary from one prompt to the next. When that happens, people look to their managers for clues: *Is this safe? Is this expected? Should I already know how to do this?*

What you signal in those moments will decide how your team approaches AI long-term, not as policy, but as culture.

1. **Name what's changing**

Start by describing the change in plain terms. Avoid the buzzwords and focus on what's actually different for your team. For example:

- *We're testing AI text tools to reduce writing time, but the final version will still need human review.*
- *We can use generative AI to explore options faster, but decisions stay with our marketing and business development professionals.*

Clarity reduces noise. It also communicates that you're not expecting overnight expertise, just thoughtful engagement.

2. **Separate risk from experimentation**

When everything new feels risky, progress stops. Leaders can help by drawing a line between *safe tests* and *production work.*

As I suggested in the last chapter, you can create a space, whether digital or physical, where staff can try new tools with no penalty for messy outcomes. Set short, written boundaries that define what they can test, what data is off-limits, and how to share what they learn. A simple framework like that signals permission to explore while keeping control where it belongs.

3. **Make curiosity visible**

People take cues from what they see. When a manager experiments in the open, by talking through a prompt they tried, or sharing what didn't work for example, it normalizes learning. You don't need to be the expert in the room. You need to be *the first to admit you're still learning.* That single gesture turns uncertainty from a private worry into a shared project.

4. **Ask for evidence, not enthusiasm**

AI adoption often swings between over-excitement and avoidance. A steady approach starts by asking for evidence, not opinions. When someone proposes applying AI to a workflow, ask:

- What problem will it solve?
- How will we measure the improvement?
- What risks could it create if we're wrong?

The goal isn't to slow enthusiasm. It's to keep ideas accountable to results, the same way any engineering or design decision should be.

5. **Bring new standards early**

AI doesn't replace professional judgment, but it does need accompanying rules. Leaders can draft simple checklists reflecting what *"responsible use"* means for their teams. These can include source verification, client confidentiality, and transparency about machine-generated material to name a few. By writing this down before issues arise, you turn potential debate into shared reference. That steadiness lets teams focus on the creative side instead of unnecessary debate.

6. **Keep communication short and regular**

Uncertainty grows in silence. Short updates, monthly or even weekly, about where the firm stands on AI and the steps you're taking towards understanding and implementation, are more effective than a single big announcement.

A few sentences from leadership *("We're seeing progress here; we're cautious here")* go a long way toward keeping speculation in check. People relax when they have a working sense of direction.

The core principle

Leadership in fast-changing environments is most effective when it looks less like authority and more like coordination. Your job isn't to know everything first. It's to create conditions where learning can scale safely.

When teams see that your stance is measured, transparent, and grounded in evidence, they'll match it. That's how uncertainty turns from a risk into part of the firm's learning rhythm.

Next, we'll look at the opposite side of this balance. Because too much reliance on structure can slow necessary adaptation. Let's explore how to ensure the controls that keep your risk reasonable don't harden into resistance.

***

Avoiding the over-correction

Once an organization begins to take AI seriously, the next common risk is over-correcting. It builds so many guardrails that progress slows to a crawl. Caution is healthy, but bureaucracy is not. Too much structure can turn a learning opportunity into another compliance exercise, and innovation drifts back underground.

1. **The pattern of over-correction**

It usually starts with good intent. Someone raises a valid concern about accuracy, privacy, or bias. Policies are drafted to reduce risk, and early pilots are paused while management "sets standards."

Months later, the team still hasn't run another experiment. The excitement that sparked the initiative fades, replaced by procedural fatigue. When that happens, AI stops being a capability and starts being a category of paperwork.

2. **Why it happens**

In technical environments, risk control feels familiar. We're used to approval chains, checklists, and designs that aren't released until they've been checked and double-checked. That mindset works for production systems and structural calculations. But experimentation only thrives when it lives by different rules.

Pilot projects need the freedom to test boundaries within known limits. They depend on iteration and feedback loops that rigid processes can't always accommodate. Without a dual structure, with one for exploration and one for production, firms end up treating every idea like it's already client-facing. Nothing gains momentum.

3. **Balancing discipline and discovery**

There's a middle ground worth protecting. It starts by defining where rigor is essential and where flexibility adds value. A simple framework is to separate *"sandbox"* work from *"delivery"* work:

Phase	Primary Goal	Review Level	Typical Output
Sandbox	Learn what the tool can do	Peer review within team	Internal notes, prototypes, demonstrations
Delivery	Apply proven methods to real projects	Standard QA and client review	Official communications, proposals, design deliverables

This division lets technical staff explore new tools with permission, not apology. Everyone knows when they're experimenting and when they're releasing work that represents the firm.

4. **Keep the rules proportionate**

A good test for any policy is whether it still encourages improvement. If every small trial requires executive approval, the policy is too heavy. If people hesitate to mention a new tool for fear of "using it wrong," the rule has become a deterrent.

Guidelines should scale with risk. It's sensible to have light processes for low-impact experiments and full review for anything leaving the building. Measured this way, governance supports progress instead of blocking it.

5. **Watch how people behave, not just what they report**

While surveys and status updates can tell one story, behavior tells another. When staff keep AI use quiet or shift it to personal devices, that usually means they're trying to innovate in spite of barriers. That's valuable feedback. It means curiosity is alive, but the structure around it needs adjustment. The fix isn't discipline.

It's dialogue. Ask what they're trying to achieve and how to make that exploration safe and visible.

6. **Leadership's signal**

Leaders set the tone for realism. A steady message that says, *"explore responsibly, share what you learn, bring concerns early,"* prevents the extremes of reckless speed and paralyzing caution. But, the example matters more than the policy. Once people see a balanced approach modeled from the top, they'll learn how to match it in their own work.

The core principle

Good governance protects experimentation; it doesn't replace it. Policies should evolve as capabilities grow, not the other way around. When a firm can adjust its rules without losing its nerve, it's ready to move from pilot projects to everyday practice.

With the cultural balance set we can turn to the mechanics of real adoption. Our next section will look at how teams can begin integrating AI into regular workflows, from documentation to project delivery, without disrupting the professionalism clients depend on.

***

Turning theory into routine

Once teams understand the potential of AI and feel safe to explore it, the real test begins. This is where you make it part of everyday marketing and business development work.

Excitement is easy in a workshop or a pilot. Turning that learning into reliable habits across pursuits, content, and client outreach takes structure and repetition. The aim isn't wholesale reinvention. It's steady integration, so AI becomes one more dependable tool in how you plan, communicate, and position your firm.

1. **Start with the everyday tasks**

Early success happens when AI clearly lightens the load. Begin where the return is immediate, and the risk is low with tasks such as:

- Drafting proposal sections or project summaries
- Pulling client intel from public sources
- Organizing pursuit notes into outline form
- Generating first-pass social captions or blog posts
- Turning meeting transcripts into follow-up points

These are time-intensive but not client-sensitive. Automating even part of them frees capacity for higher-value work, like messaging strategy or relationship building.

2. **Identify where it adds clarity**

Before changing how the team operates, map out a typical pursuit cycle or marketing campaign. Then ask:

- *Where does information pile up faster than we can analyze it?*
- *Where are we reinventing the wheel each time?*
- *Which steps depend on repetitive summarizing or formatting?*

Those are the right insertion points for AI. You're not redesigning the process. You're focusing it.

3. **Name owners and champions**

Each active marketing or BD area, including proposals, CRM, social content, and thought leadership, should have a point person who tests new tools, documents what works, and reports to the team once a month. The point isn't to create a new role. It's to make AI use accountable to results. When ownership is explicit, momentum doesn't fizzle between meetings.

4. **Create lightweight standards**

When an AI method proves useful, write it down in simple language. You don't need a policy binder, just a short note that explains:

- **Purpose:** what problem this solves
- **Tool:** which platform is appropriate and approved
- **Inputs:** what's safe to include and what's off-limits
- **Review:** who signs off before client use

These quick references will keep quality stable without smothering initiative. They also give new team members a safe place to start.

5. **Make sharing routine**

AI adoption matures fastest when learning spreads sideways. Set up a short, predictable way for marketers and BD staff to share what they've tried, such as a ten-minute agenda slot, a shared Teams channel, or a quick round-up email. Encourage quick examples instead of elaborate presentations. Little exchanges like *"It saved me forty minutes on research,"* or "*It helped tighten a boilerplate paragraph,"* keep progress visible and grounded.

6. **Translate wins into business value**

Executives and principals want to see clear outcomes, not novelty. When you report back, translate AI's effects into metrics they recognize:

- Faster turnaround on proposal drafts
- More accurate client briefs
- Fewer errors in contact lists
- Higher open rates in campaigns

Turning an "AI trial" into measurable business improvement signals that the team is investing, not experimenting.

7. **Keep your toolset lean**

Marketing software can multiply quickly, as each platform promises unique insight or automation. Once each quarter, review your active subscriptions and decide which tools are actually delivering value. A smaller, well-understood suite will outperform a stack of half-used apps every time. Simplicity scales better than novelty.

The core principle

Routine is where innovation proves itself. AI should serve the daily reality of marketing and business development by helping teams think faster, write smarter, and respond sooner. It should not change what professional communication is supposed to do, which is communicate value and build trust.

When these tools blend into the rhythm of pursuits and planning, they stop looking experimental. They just look useful.

With AI becoming part of regular marketing work, the next focus is growth. This is how you can keep teams developing new skills as both the tools and the market keep evolving. Our next section looks at how to sustain learning and retain flexibility, so capability grows as fast as technology does.

***

Building adaptive teams

Once AI is part of everyday marketing and business development, the challenge shifts from implementation to maintenance. How do you keep people learning when the tools, prompts, and platforms change every quarter? How do you preserve momentum without burning people out on "innovation fatigue?"

The answer lies in building teams that are *deliberately adaptive*. In other words, they're organized to keep learning at the same pace the environment evolves.

1. **Treat learning as a standing function**

In most firms, business development and marketing already have standing meetings, pursuit reviews, and pipeline check-ins. AI needs the same kind of cadence, but lighter.

Establish a short, recurring session (even fifteen minutes monthly) focused only on what's new, what's working, and what needs caution. This rhythm puts improvement into the schedule, not on the margins, and treats it as part of the work, rather than a detour from it.

2. **Refresh roles with learning ownership**

In adaptive teams, *everyone* carries some responsibility for insight, not just "the AI person." Writers share updates on prompt templates. Coordinators report on research tools. CRM managers flag new integrations. When each role owns a slice of the experimentation, the team evolves without any single person having to steer all of it. Knowledge travels horizontally instead of bottlenecking.

3. **Capture what you learn, lightly**

Unrecorded discovery disappears. Adaptive teams keep brief, searchable notes about what worked, what failed, and what still needs testing. This doesn't need to be polished documentation, just two or three lines in a shared document is enough. It's memory in motion, the running log that saves future time and prevents repeated mistakes. Over a year, that living record becomes a proprietary playbook built from your own experiments.

4. **Reward practical curiosity**

Curiosity is often treated as a personal trait, but it's also a management lever. When recognizing contributions, highlight the *usefulness* of an experiment, not its novelty. For example, *"Thanks for finding that prompt that cut proposal drafting time by 20%,"* is much more effective than *"Thanks for diving into the latest AI trend."*

That subtle difference teaches teams that curiosity connected to outcomes is valued currency.

5. **Build a safe channel for mistakes**

Adaptation slows when people fear embarrassment. Create a private, judgment-free place where staff can share AI attempts that go sideways—a tone mismatch in a press release draft, a mis-summarized client brief, an inaccurate reference. (Some of my funniest writing moments have come from my personal AI blooper reel.)

Instead of hiding the failure, capture what was learned. Every misstep can become a small case study in what to adjust next time. Embracing this approach reflects a broader business truth: the firms that improve fastest are the ones that fail fast and fail forward, treating each wrong turn as actionable data rather than wasted effort. When learning from the miss is built into the process, even stumbles move the organization ahead.

6. **Keep leadership visible but hands-off**

Leadership's role is to reinforce purpose, not micromanage technique. Set expectations, including responsible use, data integrity, and consistency of brand voice, then let the teams decide how to meet them. Check in periodically but not to grade performance. Instead, leadership's role is to remove barriers. When leaders model trust and curiosity, the learning culture replicates itself naturally.

7. **Manage fatigue before it manages you**

New tools arrive faster than people can evaluate them. A good adaptive team knows when to say *enough for now.* Institute a simple rule: new tool reviews happen on a schedule (for instance, twice a year), not every time a shiny object appears. This reduces noise and keeps attention on getting the most from what's already approved. Saying "not yet" to endless novelty is part of long-term adaptability.

The core principle

Adaptability isn't a one-time reset; it's a practiced behavior. Teams that grow through change treat learning as routine, curiosity as accountable, and mistakes as data. That steady, structured, and unemotional discipline is what converts disruption into ongoing advantage.

With an adaptive team culture in place, the next step is scaling that maturity across the firm as it expands from individual teams to enterprise-wide communication and brand integrity. So, let's explore how firms present their AI capability to clients and the market without overpromising or feeding the hype cycle.

***

Aligning communication and reputation

As AI becomes part of daily work, an external question will quickly follow: *How do we talk about this responsibly?*

In professional services, your reputation is built on trust, not trendiness. The firms that navigate this moment best are the ones who describe their use of AI with clarity, neither exaggerating capability nor downplaying progress.

1. **Don't promise magic. Describe method**

Clients don't expect perfection; they expect competence. Focus on *how* AI supports your work rather than implying that it *does* the work. Replace phrases like *"AI-powered proposals"* or *"AI-driven insights"* with statements that show stewardship and integration: *"Our marketing team uses AI tools to improve response time and consistency while maintaining full human oversight."* That framing keeps ownership with your professionals, where it belongs. And it positions AI as disciplined support, not instant genius.

2. **Translate backstage improvements into client confidence**

   Most AI benefits are invisible, such as faster background research, tighter scheduling, and cleaner data. But *process transparency* builds credibility. For example:

- If AI helps your team synthesize client data, tell clients that those summaries allow your firm to respond sooner and more accurately.
- If it automates routine proposal sections, explain that more staff time is available for tailored strategy and design emphasis.

By looping client benefit into every mention of AI, you protect your reputation while demonstrating competence.

3. **Use consistent language throughout the firm**

   Marketing, BD, and leadership voices often describe innovation with different vocabularies. Instead, establish a shared definition of what "responsible AI use" means for your firm, where it belongs, who manages it, and what internal checks exist. When that phrasing appears consistently in proposals, interviews, press releases, and social channels, it communicates maturity. Consistency is a quiet signal of credibility.

4. **Define your learning stance publicly**

   Adaptive firms don't wait until they have everything perfected. They communicate *in motion.* That doesn't mean oversharing experiments. It means being comfortable saying, *"We're learning where AI adds value and where it doesn't. Our priority is using it responsibly, with human review on everything we deliver."*

   This approach combines humility and confidence. It tells the market you're progressing thoughtfully. It's a professional echo of the same "fail-forward" mindset we talked about using internally. Learn openly, correct quickly, and move steadily.

5. **Guard brand integrity while testing new voices**

AI can generate tone, rhythm, and phrasing, but only people can sense whether the message *feels like you.* Before publishing AI-assisted content externally, run it through the same brand filters you use for any marketing piece: tone, accuracy, alignment with firm values. Clients hire competence, not flash. Discipline in communication is how you keep trust through every wave of new tools.

6. **Let proof accumulate quietly**

The firms who ultimately lead these transitions don't brag about being early adopters. They earn the label through outcomes. A year from now, what matters won't be *that* you used AI, but *what changed because you did*:

- Shorter proposal cycles
- Better client engagement quality
- Stronger hit rates

Speak less about the tool and more about the results. Let your success make the announcement.

7. **Integrate lessons into your brand story**

Every learning cycle and every small "fail forward" moment adds depth to your narrative of competence. When you share stories of refinement *("we discovered that this approach works better than that one")*, you paint a picture of a firm that learns faster and adapts smarter. Over time, that becomes part of your brand equity. It's not perfection, but professionalism in evolution.

The core principle

Reputation in the AE professions isn't built on being first. It's built on being *reliably better*. Communicating about AI should reinforce your steady improvement, accountable learning, and visible human judgment. Firms that master that balance will be trusted to use new tools long after the excitement fades.

With a grounded communication strategy in place, the learning curve begins to level out. The fears that opened this chapter—of being replaced, left behind, or overtaken by hype—have now been reframed as signals to guide wiser action.

Before moving forward, it's worth pausing to take stock regarding any shifts that have occurred in your own perspective. The reflections that follow invite you to capture that change before we turn toward building on that clarity in the chapters ahead.

***

Reflection exercise: Taking stock before you move forward
Purpose: To help you pause and locate yourself in this transition: from uncertainty to understanding and from hype to habit. This reflection isn't a test; it's a quick inventory. Use it privately or as a prompt for conversation with your marketing or BD team.

1. **One fear**

   What hesitation still sits under the surface when you think about AI in your marketing, business development, or sales work?

   Example: "I'm worried we'll lose the personal tone that sets our proposals apart."

   Reframe it: What would it look like to turn that same fear into curiosity or a learning goal? *"I want to test how we can preserve voice and relationship-building as we automate."*

2. **One piece of hype you've outgrown**

   What promise or buzz have you stopped believing?

   Example: AI will instantly write perfect content or deliver automatic client insights.

Reframe it: What recent experience showed you the real, smaller, more useful advantage instead?

3. **One opportunity you can act on**
   What small action could you take this quarter that moves AI from concept to routine?
Examples:

- Pilot AI for first-draft proposal sections.
- Use it to summarize client meeting notes before pursuit strategy sessions.
  Keep it modest. The goal isn't scale; it's momentum.

4. **One lesson to share**
   Who else in your firm could benefit from what you've learned so far? Write one sentence you could send today, such as a tip, observation, or small success to model that "fail-forward" habit of open learning.

Remember: Reflection isn't a pause in your progress; it *is* your progress. Today's clarity gives strong purpose to your later experimentation.

Chapter reflection: Fear, hype, and opportunity

When this chapter began, the conversation around AI felt loud, emotional, and uncertain. Now it should feel more familiar; a leadership topic grounded in professional habits rather than technological panic. We've reframed fear as attentiveness, hype as early-stage confusion, and opportunity as the steady work of refinement. In every example and beat, one pattern has held true: progress favors those who stay curious, communicate clearly, and give their teams space to learn in public.

Each firm will adapt at its own pace, but the direction is universal—toward *measured adoption, transparent communication, and resilience through learning.* The same temperament that carried your firm through CAD, BIM, CRM, and every other so-called revolution will carry you again. What emerges from this chapter isn't certainty, it's steadiness. Steadiness to *test, learn, and adjust* and to fail forward until new tools become trusted tools.

With that steadiness in place, the next step is outward as you turn internal competence into visible differentiation. In Chapter 4, we're going to explore how to translate your firm's AI experience into a credible market narrative, so clients, partners, and competitors see not just that you're keeping up, but that you're moving with purpose.

CHAPTER 4

The smart firm formula: Human + Machine = Momentum

Rethinking 'Smart'

"Work smarter, not harder," has been the mantra for decades. But when firms first talk about "getting smarter," the conversation often drifts to software lists. Someone mentions a new CRM plug-in, another talks about generative proposal tools, and before long there's a chart comparing features instead of a plan connecting people.

"Smart," however, was never about owning more tools. It's about ensuring every person in the firm can see the same picture and move in the same direction. Smart firms aren't defined by their technology stack; they're defined by how clearly their people connect information, intention, and action.

The misunderstanding of "smart"

Over the past decade, the world has somehow taught us to equate *smart* with *automated.* Smart phones, smart cars, smart buildings… these are all shorthand for systems that act on their own. But professional services don't work that way. Momentum in our arena depends on collaboration, persuasion, and trust. Nothing closes a client pursuit by itself.

Which means our definition of *smart* has to change.

A smart firm isn't one that replaces human effort. It's one that aligns *human insight with machine assistance* so the insight scales, and the assistance stays grounded in judgment. In other words:

Human + Machine = Momentum.

That's not a slogan. It's a filter: a fast way to test whether an AI move will truly create value.

- If the machine saves time but confuses communication, there is no momentum.
- If the human adds context but the data sits siloed, there is no momentum.

Momentum only appears when both parts push in the same direction.

Where momentum comes from

Picture a pursuit team heading into interview prep. The AI summarizer scans project archives and pulls past Q&A transcripts allowing a marketer to distill patterns in client concerns from those summaries. The team walks into rehearsal with clarity and confidence.

That is *human + machine.*

Now picture another team that spent three hours polishing automatically generated slides but never discussing narrative flow. That is *machine – human,* and it shows.

Momentum isn't about speed; it's about coherence. When everyone can see why the data matters, decisions compound faster, and success compounds with them.

The new meaning of "smart"

So, let's redefine the word for this profession: *Smart is synchronization.*

It's the firm where data, people, and stories run on the same track. It's where automation serves awareness, and awareness fuels better work.

That's what we're going to unpack next. We'll explore how "smart" behavior shows up in three dimensions: *seeing, deciding, and communicating.* And we'll explore how to use them to create true momentum inside your firm.

But first, let's pause for a quick mental check:

- Does "smart" at your firm still mean shiny and new, or coordinated and clear?
- Are your best people learning from your machines? Or are they working around them?

What you notice in those answers is where your next acceleration begins.

The three dimensions of 'smart'

When you strip away the jargon, every "smart" behavior inside a firm comes down to three verbs: *see, decide, and communicate.* Each represents a partnership between human intuition and machine precision. When done well, this handoff generates momentum that's visible to the whole team.

1. Seeing: Turning information into insight

A smart firm *sees* better than its competitors. Not because it has access to more data, but because it knows what to look for and how to connect the dots.

- *The human strength:* curiosity and empathy. People can sense the *why* behind a client's pattern, not just the *what.*
- *The machine edge:* range and recall. AI can scan thousands of documents, posts, or proposal win-loss notes in seconds, surfacing signals that the human eye alone would miss.

In practice:

Dylan is a BD manager at a small firm in the Northeast. Before pursuing a large institutional project, she had AI comb through partnership lists, media releases, LinkedIn updates, and recent RFP language for any patterns in the client's decision criteria. The machine reported that, in the past year, sustainability metrics appeared twice as often as cost efficiency.

This led her to conclude that the client was reframing value to include long-term environmental return and she adjusted the pitch deck to lead with lifecycle value, not bid price.

That's how you see with context.

2. **Deciding: Pairing judgment and speed**

Smart firms decide faster without cutting corners. They use AI as a clarity tool, not a shortcut.

- *The human strength:* judgment and ethical sense. People understand nuance, politics, and trust, which are things algorithms can't measure.
- *The machine edge:* speed, memory, and unemotional pattern detection. AI can compare historical pursuits, filter by budget, geography, or client type, and highlight which types of pursuits statistically favor your firm's strengths.

In practice:

The pursuit team at a 50-person firm in Texas input project data into an internal AI model that was prompted to grade opportunity fit based on prior performance factors. The machine produces a 72% match score. The team didn't treat that as gospel, but let it become a conversation starter, asking if that aligned with their strategic focus? Did they have a story strong enough to raise that 72 to a 90?

In this case, the machine accelerated clarity and removed bias from the Go/No Go decision process. The people kept the decision ethical and strategic.

Momentum, again, came from connection.

3. **Communicating: Using machines to personalize, not automate**

While seeing and deciding aren't evident to your client, communication is highly visible. This is where the AI stories of many firms drift off course. They use tools to mass-produce language instead of to mass-customize it.

- *The human strength:* voice, tone, and emotional calibration. Professionals understand when to sound confident, empathetic, or conversational.
- *The machine edge:* scale and consistency. AI can draft, rephrase, or check alignment with branding and readability standards.

In practice:

A marketing coordinator in a large firm in Florida had to draft a proposal intro. He used AI to generate three different tonal variations and it produced formal, concise, and relational versions of the text in a matter of seconds. He decided which tone was most appropriate for the client, and then re-edited the winner until it sounded like the firm's actual voice instead of "AI-neutral English." What the client received felt personal and poised.

The machine expanded the options, and the person made them authentic. That's the communication balance every client now expects: timely *and* trustworthy.

Why these dimensions matter

Every firm already performs the three daily actions of seeing, deciding, and communicating. The difference between "busy" and "smart" is *how visible and connected those actions are across people and platforms.* When each function operates in a vacuum, the firm lurches forward in short bursts. When data flows between them, while managed ethically and interpreted wisely, the organization stops sprinting and starts gliding.

You can measure that shift by perception as meetings feel shorter, pursuits feel clearer, and cross-functional email threads shrink. That's momentum at work.

Looking ahead

In the next section, we'll unpack what this *Human + Machine = Momentum* model looks like in practice. You'll see how small operational alignments, such as data hygiene, templates, and internal language, translate the theory into daily progress. Because knowing the three dimensions is one thing. Building them into the firm's rhythm is where the true "smart" begins.

***

The smart firm equation in practice

Human + Machine = Momentum shouldn't just be a slogan. It should be a working formula. The question is how to make it tangible inside your firm. How do you know whether you're already "smart," still manual, or halfway in-between?

Most firms don't need a new department or a grand rollout. They need a visible model that shows how your alignment with these powerful tools grows over time.

The three stages towards a smart firm

Think of it as a curve towards maturity. Every firm travels it, even if informally. The goal isn't to rush, but to move *deliberately* from one stage to the next.

Stage 1: The manual foundation

Hallmark: Human effort carries everything.
Feels like: Dependable, but exhausting.
AI role: Minimal or reactive. Perhaps there's a single enthusiast testing tools in isolation.

At this stage:

- Insight lives in people's heads or in spreadsheets that haven't been shared.
- Decisions rely on experience more than data.
- Communication shifts with whoever's doing the writing that week.

Momentum in this phase depends on personal heroics. Success happens, but it's slow and inconsistent.

Healthy behaviors: loyalty, craftsmanship, sound relationships.
Risk: burnout and lost opportunity knowledge if key people leave.

The takeaway: While potentially frustrating, especially to the go-getters and enthusiasts, this manual foundation isn't failure, it's the soil from which "smart" can grow. But it can't be a permanent state because it can't sustain scale.

Stage 2: Augmented workflows

Hallmark: The human lead is intact, but technology is now a recognized partner.
Feels like: Relief.
AI role: Point solutions in which targeted helpers support specific tasks.

At this stage:

- Teams use AI tools to draft text, summarize meetings, refresh templates, or analyze pursuit data.
- Workflows get measurably faster, but parts still operate separately.
- The firm begins to create internal norms for review, editing, and ethical disclosure.

Healthy behaviors: Curiosity, experimentation, sharing wins and fails publicly.
Risk: Scattered adoption with pockets of efficiency but no firm-wide learning.

Momentum in this phase is visible, but uneven. Each success story becomes a seed for the next level. But only if someone connects the dots.

Stage 3: Connected intelligence

Hallmark: AI capabilities are woven into the firm's collective rhythm.
Feels like: Clarity and composure.
AI role: Integration, in which systems and habits are working in concert.

At this stage:

- Research, pursuit strategy, and proposal production draw from shared data sources.
- Decisions reference transparent metrics, not hunches alone.
- Internal knowledge flows freely between marketing, BD, and operations.

Healthy behaviors: Communication fluency, visible trust, and cross-disciplinary learning.

Risk: Complacency, or believing "we're done" when maintenance must become the new skill.

Connected intelligence isn't the end. It's the point where "AI" disappears into normal work. The focus shifts from adoption to adaptation.

Putting the model to work

Use this three-stage model not as a grading chart, but as a conversation starter:

- *Ask* each team to identify which stage best describes their current workflow.
- *Discuss* what would make the next step *practical* instead of merely aspirational.
- *Record* one small experiment that tests a bridge from today's stage to tomorrow's.

For example: If proposal content is still manually curated (Stage 1), conduct a controlled pilot test by having AI summarize a number of past project write-ups (Stage 2). Or, if your teams already use AI for research but data lives in multiple silos (Stage 2), create a simple shared repository that updates automatically (a step toward Stage 3).

Momentum isn't built in a single leap. It accumulates through iteration with small, recoverable experiments that teach faster than they break.

Why this equation works

The beauty of *Human + Machine = Momentum* is that it honors both halves of the equation and their respective contributions to your firm. Humans remain the source of vision, creativity, and judgment. Machines extend that energy outward while performing faster, further, and with fewer dropped handoffs.

Every success in modern BD and marketing rests on that interplay:

- When tools clarify rather than complicate, people can lead more effectively.
- When people trust what tools can and can't do, they spend more time winning work instead of chasing "what's next."

This is what makes AI a cultural advantage, not just an operational one. The equation turns abstract technology into visible teamwork.

Looking ahead

In the next section, we'll dig into what sustains this alignment once it starts to work. We'll see how trust, language, and leadership behavior determine whether your firm's momentum doubles or dissipates.

This math only holds if the culture supports the equation. Otherwise, even the smartest tools will coast to a stop.

***

Culture as the multiplier

If the earlier sections have given us the mechanics of the *Human + Machine = Momentum* equation, we now need to look at what keeps it alive. Because the real multiplier isn't the math. It's the culture around it.

Every firm, no doubt yours included, has felt this truth in some form.

I've witnessed this difference time and time again as firms work to integrate new processes, whether they've been driven by technology, market shifts, or leadership change. One group adopts new tools and finds its stride as people share tips, small successes build, and what began as an experiment soon becomes everyday practice. Another group has the same tools but none of that lift. Energy fades. Curiosity gets replaced by caution. It's the same machine and the same intention, but two different outcomes.

The difference is trust.

Momentum grows where people feel safe to learn

When change enters the room—and *all* change feels scary to one degree or another—people scan for cues. If the signal from leadership says, *"this is a safe place to figure it out,"* experimentation begins. If the signal sounds more like *"don't mess this up,"* curiosity shuts down.

Culture multiplies progress because it determines whether people share discoveries or hide them until they're "perfect." A smart firm would rather have ten small, visible lessons than one hidden "success" that no one else learns from.

Leaders will set that tone in one direction or the other. And usually without realizing it. The simplest way to encourage confidence is to narrate your own learning out loud: "*I tried this tool for our project summary. It saved time, but I didn't like the tone. Let's see if we can teach it to match our voice.*"

That single sentence gives everyone permission to try, observe, and adjust. It also reminds the team that AI isn't judging them; it's joining them.

Language matters more than policy

Policies around ethics, data protection, disclosure, and other concerns are necessary. But the daily culture of your firm lives in the words you choose. Saying *"Let's test this"* feels very different from *"Don't get that wrong."* Saying *"We're exploring how this helps us serve clients better"* keeps the emphasis on purpose, not performance.

In every firm that makes a smooth digital turn, you'll hear the same tone of measured curiosity in conversations. It's a confidence that doesn't over-promise but doesn't resist progress either. People begin to describe their work in terms of *partnership* with clients, colleagues, and even the tools themselves.

In the very creation of this book, in which I've made extensive use of AI as a writing partner, I've come to know and appreciate my co-writer as a valued collaborator. We have actual conversations and even debates about the best way forward. We discuss goals, evaluate output, and test different approaches in the identical manner I would if I were working with a human partner.

After almost a year of working together, what feels like the human equivalent of mutual trust and respect has grown, making my job infinitely easier and more pleasant.

When you share your experience with the others in your firm, that tone of confidence and success spreads faster than any training program. Culture grows by imitation.

Momentum feels like permission

Ask around in a firm that's finding its stride and you'll hear a different kind of energy. There's a little more laughter in meetings. Deadlines still exist, but the air feels lighter. People aren't tiptoeing around the technology anymore. They're inviting it in, deciding together when it belongs at the table and when it doesn't.

That's what I mean by culture as the multiplier: AI doesn't make that atmosphere; it amplifies whatever already lives there. If people feel respected, AI speeds up respect. If they feel uncertain or competitive, it speeds up that, too. So, the work of leaders is to build the environment where clarity, empathy, and kindness move faster than anxiety.

The equation's little secret

You can buy technology. You cannot buy trust. And trust is what compounds every other investment.

When your team believes that learning together is valued more than looking flawless, they start pushing past the awkward first drafts, the half-useful results, the nervous laughter during demos, and the outright laughter at the hallucinations. That's the phase where culture either multiplies momentum or drains it.

The leaders who manage this moment best are the ones who stay human in public. They'll laugh when something odd happens, they thank people who raise problems, and they keep the tone steady: *"This is new territory, and we're walking it together."*

Looking ahead

In the next section, we'll take that spirit of trust and turn it into something practical: a reflection exercise that helps you gauge whether momentum is already taking hold in your own culture. Because once people feel safe to learn, the next question becomes simple: *How do we help them see the progress they're already making?*

***

Your momentum checklist

If you've made it this far, you've already done the hard work. You've redefined "smart," you've seen the pattern, and you've started to feel what momentum looks like when it shows up.

This last section of Part 1 in the book is simply a mirror, an opportunity to look at the state of AI use and integration in your own firm to notice what's already working, and what might need a nudge. Think of it less as a checklist and more as a gentle pulse-check. No grades. No right or wrong. Just questions that tell you where energy is naturally moving, and where it's getting stuck.

1. **Look for signs of momentum**

Ask yourself, or better yet, ask your team, a few simple questions:

- When someone experiments with a new tool, do others ask, *"Will you show me how you did that?"*
- Are small wins shared out loud, even when they're imperfect?
- Does the pace of meetings feel a little lighter, not because there's less work, but because the work connects more easily?
- Do you hear people using words like *we*, *learned*, or *tried* more often than *I*, *fixed*, or *should*?

If you recognized even one of those in your own environment, momentum has already started. It may be quiet, but it's real.

2. **Notice where the energy hesitates**

Momentum doesn't vanish overnight. But it will hesitate when fear or fatigue creep in. Be curious about those slow spots:

- Do certain teams still treat AI like something they have to "get permission" to use?
- Are other teams building walls and hanging "No AI!" signs on the doors?
- Is feedback mostly silent, waiting for an expert to declare what's right?
- Are people so polite about avoiding mistakes that they've stopped talking about what they're discovering?

These are not faults; they're signals. They point to the very places where trust is asking for reinforcement. Every hesitant moment is a conversation waiting to happen. It's a chance to remind everyone that learning *in* public is part of the job now, not a side project.

3. **Choose one small experiment**

Momentum builds one small step at a time. Choose a single area where human energy and machine help could meet a little more elegantly this month. Something light, reversible, and visible.

- Ask a junior marketer to compare two AI-drafted email intros and explain which sounds more authentically "you."

- Capture lessons learned after a pitch and ask AI to summarize themes for future reference.

Whatever the act, keep it small enough that success looks like *insight*, not *perfection.*

4. Talk about what you see

The fastest way to lock in progress is to narrate it. Gather the team, and simply say, *"Here's what got easier this month. Here's what still felt clumsy. What should we try next?"*

That rhythm of *observe, adjust, and repeat* is how curiosity turns into organizational confidence. It's also how "AI adoption" stops feeling like a project and starts feeling like culture.

Closing reflection: What momentum feels like

By now, you've probably noticed that momentum doesn't announce itself. It hums. It's the conversation that moves more smoothly than it did last quarter. It's the proposal draft that came together without panic. It's the subtle confidence that the humans are still in charge and the machines are finally earning their keep. When that starts to happen, pause and name it. Tell your team, *"This is what progress feels like."* Because momentum, once recognized, tends to stick around.

Looking ahead

And now we're ready to move on from *Part 1: The Rewire: Mindset Before Mechanics.* You've established the mindset that makes every tool worthwhile: a culture of trust, curiosity, and calm experimentation.

In *Part 2: Smarter Intelligence*, we'll carry that mindset outward while discovering how to see clients more clearly, anticipate their needs, and build relationships that feel both personal and scalable. Because the real promise of AI isn't that it makes us faster. It's that it gives us back the time and the confidence to be more human where it matters most.

PART 2

SMARTER INTELLIGENCE: FROM INSIGHT TO EXPRESSION

CHAPTER 5

Instant Insight: Research in Minutes, Not Days

Before and after

Michelle is a Marketing Coordinator for a small architectural firm in the Midwest. One Tuesday afternoon, she sat facing a cascade of browser tabs that multiplied like rabbits. She was three hours into what she liked to call *"client reconnaissance,"* the pre-proposal ritual of scouring websites, news articles, and annual reports for anything that might help personalize the pursuit.

Her notebook held a patchwork of bullet points that included a board appointment announcement, the client's recent award, and a half-remembered quote from an interview she couldn't relocate. Somewhere in that tangle, an angle was supposed to emerge. But it didn't.

By five o'clock she had a headache, a half-formed theme about "community engagement," and an untouched slide deck. The team meeting was set for 8 a.m. the next morning. She closed her laptop and told herself, *"At least I tried."*

Fast-forward six months. Same Michelle. Same kind of pursuit. Different Tuesday. Only this time, she began not with a search engine, but with a short, purpose-built prompt in her AI workspace:

"Summarize the last 18 months of public news and statements from <client name>. Highlight recurring priorities or phrases that reveal what success means to them."

In less than five minutes her screen was filled with an organized digest:

- *Recurring themes:* "resilient design," "long-term community value," "ethical partnerships."
- *Notable language shifts:* sustainability → *resilience*, supplier → *partner.*
- *Quote frequency:* CEO referenced "future neighbors" five times in three interviews.

While her summary wasn't perfect (most AI summaries aren't) it was a map and a distillation that let her *see the pattern sooner and more clearly.*

She cross-checked a couple of points, added context from her own experience with the client sector, and by 9:30 she'd drafted an opening paragraph that spoke the client's language, not boilerplate.

When her team reviewed the draft, someone asked, "How did you find that phrase they keep using—*future neighbors*?" Michelle replied. "I didn't find it. I asked for it."

That's the inflection point we're going to explore in this chapter. It's the shift from hunting for information to shaping the insight you gain from the information.

In the old model, professionals survived by persistence. We manually assembled whatever fragments we could find in the hopes that meaning would appear. In the new one, insight arrives much earlier because the machine pre-organizes what the human mind is best at interpreting. The difference isn't just five minutes versus three hours. It's *clarity versus clutter.* And as we'll see, that small shift that's repeated pursuit after pursuit will change the texture of your entire workweek.

AI doesn't find truth. It finds terrain.

After a story like Michelle's, the temptation is to believe the magic lives inside the model. It doesn't. That five-minute summary didn't hand her the truth about the client; it simply cleared the fog so she could *see where to look next.* That's the real distinction that design firm marketers have to realize: *AI isn't a compass that points to truth. It's a terrain map, or a quick sketch of the landscape that lets you decide which hill to climb.*

When we confuse the two, we get lazy. We mistake proximity for accuracy. We quote summaries as if they were sources. We start believing that speed equals certainty. And that's when the quality of judgment, which is the one thing machines don't possess, starts to erode.

In working practice, the best researchers treat AI output like aerial photography. It's a wide, fast view that still needs someone on the ground to walk the site. The machine can show you the rivers and ridges, but it can't tell you which valley hides the water source, or which word the client used that hides a political nuance.

So, what is AI research good at?

- *Surfacing the obvious faster.* It eliminates the drudgery of ninety clicks.
- *Noticing repetition.* It highlights phrases or topics that human bias might overlook.
- *Connecting dispersed signals.* It can draw a thread between a quarterly report and a social-media comment thread that no human would have time to cross-reference.

And what still belongs squarely in your hands?

- *Interpreting motive.* Machines can't feel why a phrase matters to this client, this year.
- *Testing reliability.* Only a human can decide which sources merit belief.
- *Applying context.* The same data means something different in a hospital pursuit than in a university one.

That's where the balance lives in the Human + Machine formula we talked about in Chapter 4. The tool lays out the landscape and the professional decides where the meaning lives. The subtle advantage is not that you finish faster, but that you start smarter.

Here's a practical test you can build into your next research sprint:

1. Let the AI collect and summarize.
2. Before you edit a word, ask yourself, *"What is this actually showing me?"*
3. Then re-label the output in your own words: *pattern, signal, hunch, or narrative.*

If you can translate the output into a human hypothesis, you're doing research. If you can't, you're just cataloguing.

The firms that master this distinction never fall for the illusion of certainty. They use machines not as truth-makers but as *truth-accelerators.* As tools that expand curiosity rather than replace it. That's the mindset shift that unlocks every tactic we'll talk about in this chapter.

The three levels of insight: Signals, patterns, and meaning

Once you stop expecting truth on a silver platter, you can start noticing how insight actually develops. It doesn't arrive all at once, like a lightning bolt. Instead, it stacks. Each layer builds on the last until the noise starts to organize itself into something that helps you decide, not just describe.

Inside a smart research process, those layers tend to fall into three levels: *signals, patterns, and meaning.*

1. Signals: The clues

Signals are the raw fragments. They're the mentions, numbers, and phrases that keep resurfacing as you sift through material. They're the blinking lights on the dashboard that say, *"something's happening here."* A client repeats "resilient design." A budget line shifts toward community outreach. A LinkedIn post emphasizes "legacy impact."

AI excels at surfacing these quickly because it can scan across thousands of public touchpoints and tell you what shows up most often. Your job isn't to react to them, but to *collect them without judgment.* Signals are clues, not conclusions.

Human habit to keep: curiosity without confirmation. Resist the urge to decide what it all means before you see more of the puzzle.

2. **Patterns: The connections**

When several signals start repeating, they begin to form recognizable contours. Patterns show you relationships such as how one theme leads to another, or how priorities change over time.

- Maybe every mention of *resilience* appears alongside *cost predictability.*
- Maybe public statements show a pivot from *innovation* toward *inclusion.*

Those echoes are pattern territory, in other words, the middle ground where raw data starts turning into conversation material. Here, the partnership between human and machine gets tighter. AI can cluster themes or visualize frequency, but only you can ask, *"so what?"* Patterns matter only when they help you align your story with what the client values right now.

Human habit to keep: sense-making. Ask, *"If this pattern continues, what is the client likely to care about next?"*

3. **Meaning: The 'why' that moves people**

Meaning is where insight becomes advantage. It's the layer you can't delegate because it relies on emotion, ethics, and empathy. This is where you translate a pattern like "resilient design" into the human concern underneath it: *They're tired of short-term fixes.*

Machines can shovel data, but only humans can interpret aspiration. Meaning connects the external research back to your internal strategy:

- How does this align with our strengths?
- Where do we help them look good to *their* stakeholders?

- What story or question will make them feel, "*These people understand us?*"

That last one—understanding—is the goal. Because in marketing and BD, relevance is rarely about who has the most information. It's about who identifies the *truths that matter.*

When you start thinking in these three levels, you stop treating research as "gathering facts" and start treating it as "building alignment." Your AI tools help you live at Level 1 – the signals. Your analytical skills move you into Level 2 – the patterns. Your empathy and professional humanity carry you to Level 3 – the meaning. There is no plug-in app that will bridge that gap for you. But used together, the layers make research feel lighter, faster, and infinitely more useful.

Next, we'll take that framework and apply it to a real-world workflow. We'll explore how to ask questions that lead AI beyond facts into usable focus for your team's next pursuit.

***

Building a research prompt and validating the output

Theory feels elegant until you have to open a blank field and make it work. That's often where marketers stall. The idea of "AI-assisted research" sounds efficient, but the screen's cursor blinks back like a dare. What do you actually *ask* for? What words deliver usable intelligence instead of trivia?

The good news is you don't need to become a "prompt engineer." You just need to think like a marketer who knows how to frame a question.

At its heart, writing a good research prompt is the same as running a good client interview. You're defining the topic, the scope, and the outcome you hope to understand. The difference is that your listener happens to be a machine that answers literally.

Step 1: Frame the purpose, not the task

Poor prompts describe *activity*: *"Search the web for information about..."*

Good prompts describe *intent*: *"Summarize the client's public statements from the last 12 months to identify recurring priorities and the language they use to define success."*

One asks for content, the other asks for clarity. The second phrasing gives the model edges, answering questions such as what timeframe, what kind of input, and what kind of output. It signals that you're after *themes*, not a copy-paste of facts.

That simple difference in framing the prompt determines whether you get a haystack or a handful of needles.

Step 2: Narrow by context

Machines don't know what context you're operating in unless you tell them, so you need to add it.

- *"Focus on projects or initiatives relevant to higher-education clients."*
- *"Prioritize content related to healthcare facilities and long-term campus planning."*

Every extra line that defines audience, geography, or market focus increases relevance and reduces cleanup. As I suggested earlier, imagine that you're giving instructions to an intelligent, resourceful, but inexperienced intern. The more you define the 'what' and the 'why' of the task, the better they're going to perform. Think of it as putting a fence around your fieldwork. You're still gathering openly, just with boundary lines that protect meaning from unwanted "drift."

Step 3: Define deliverables you can use

A model's raw answer is rarely "presentation-ready." Guide it toward structure you can drop straight into your workflow.

"Return the results as three short bullet lists: recurring phrases, emerging priorities, and potential messaging angles for a proposal introduction."

That cue alone can save an hour of post-processing. You've told the system *how* to package its response in a format that speeds human editing rather than adds work.

Step 4: Validate before you celebrate

Even a good-looking output can mislead you. Before you include the information in any pursuit file, check three things:

1. **Source clarity: Trace, don't trust.**

- Copy one or two key lines from the AI summary and paste them into a regular search engine or your subscription database.
- Make sure at least two independent sources point to the same fact or quote.
- If you can't trace the origin in sixty seconds, treat it as a *clue*, not a *claim.*

2. **Recency: Confirm the time stamp.**

- Ask the model directly: *"What is the publication date of your most recent source?"* (Many will surface it if prompted.)
- Or run a quick manual check by searching the same topic in Google News or company press releases to confirm events fall within the last 12–18 months.
- Out-of-date data can still be useful for trend contrast. Just label it clearly as "historical context."

3. **Consistency: Cross-reference with known truths.**

- Compare the AI's main points with what your team has learned from client meetings, RFP language, or trusted industry intel.
- If the pattern aligns, confidence grows. If it contradicts lived experience, flag it for discussion before presentation.

- Use this quick double-check sequence: first confirm the AI's results with your own team's experts, then verify them with a reliable public source.

If the output passes those checks, you can treat it as *orientation data*—the early sketch of terrain we talked about. If it fails one, correct or delete it quickly. Because it's written so cleanly, model hallucination can feel very persuasive. But credibility (and your reputation) lives in confirmation, not the elegant cadence in which AI can write.

Step 5: Translate output into insight

Once you've gathered and cleaned the AI summary, start climbing the three-level ladder you built in the last section. Ask yourself:

- Which items here are *signals*—raw mentions worth tracking?
- Which combine into *patterns*—directional movement or shift?
- What tentative *meaning* could we draw, pending real-world validation?

That conversation, which is held out loud with your team or in notes to yourself, is what converts data into guidance.

The 20-minute circuit

For practical teams, here's a simple rhythm you can repeat before any significant pursuit meeting:

- ***Five minutes***: Compose or reuse a proven prompt with updated parameters.
- ***Ten minutes***: Scan and clean the AI-produced summary, verifying obvious sources.
- ***Five minutes***: Label outputs as signal/pattern/meaning and outline one hypothesis for the team to discuss.

That's it: twenty minutes from blank screen to strategic talking points. No magic, just alignment.

As this rhythm becomes habit, "research" stops being a marathon and turns into an exciting warm-up lap. You enter every meeting with shared vocabulary and early focus.

Ethical checkpoint: Confidentiality and verification

Speed can tempt even the careful professional to drift into gray zones. When data seems effortless to gather, it's easy to let your guard drop. We paste, query, and share before pausing to ask, *"Should I?"* This checkpoint isn't here to slow you down; it's here to keep your momentum trustworthy.

There are only two questions that matter:

1. *Can I use this information?*
2. *Can everyone involved see how I used it?*

And three practices that will ensure you stay inside the safe and ethical guardrails.

1. **Keep client and firm data out of public models**

Never insert internal material including draft proposals, client names tied to private discussions, or pricing data into any model that lives on public servers. If you wouldn't hand that paragraph to a journalist, don't hand it to an algorithm you don't control.

Instead:

- Use anonymized placeholders ("<Client X>") when testing prompts in open tools.
- If you're working on sensitive pursuits, confine your AI research to vetted enterprise systems or on-premise solutions with nondisclosure protections.
- When in doubt, ask your IT or legal lead before you experiment.

The guideline should be that if you're transparent first, you'll never have to apologize later.

2. **Treat AI output like an intern's first draft**

Respectful skepticism is a virtue. Everything the model gives you is editable, debatable, and subject to your signature of professionalism. That means you, and only you, own the final fact check. Simple daily habits keep this instinct sharp:

- Before quoting numbers or attributions, trace them to an identifiable public source.
- Convert every "the model said..." into a verifiable citation.
- If no source exists, restate the idea as observation, not evidence: *"Recent commentary suggests..."* rather than *"It's a fact that..."* Humility protects credibility.

3. **Clarify what's artificial and what's authentic**

Clients don't expect you to swear off technology. They just expect honesty about how you use it. If an AI application contributed to research or formatting, acknowledge that inside the team. And if the work ever appears in a public deliverable, lead with disclosure that shows professionalism, not confession: *Research summarized with the assistance of AI tools and validated by the firm's marketing team.*

A line like that turns suspicion into reassurance. It signals discernment, not dependency.

4. **Stay within the confidence corridor**

There's a simple moment that separates confident experimentation from risky behavior, and it's the pause before you hit "Send." That pause *is* the confidence corridor. It's the space where you ask, *"If someone read this tomorrow, would I be comfortable owning it?"*

Working inside that corridor doesn't slow you down; it steadies you. It keeps pace with the same values of discretion, reliability, and respect for client trust that your firm has always worked by. The technology may be new, but the ethics aren't. They're the same ones that built your reputation, now applied to a faster toolset.

When you and your team keep your vision through that lens, you're not just protecting confidentiality; you're modeling credibility. And that's what clients still buy before anything else.

A short re-grounding

Ethical research doesn't mean stopping to double-check every keystroke. It means keeping the work transparent and traceable. It is quick to explain and easy to defend. Think of these as your guardrails, light enough not to slow you, strong enough to keep you on course:

- *Ask before you share.* If the data belongs to a client or to your firm, confirm permission first.
- *Verify before you cite.* Every line earns a source before it earns a slide.
- *Disclose before you're asked.* Transparency reads as confidence, not caution.

Follow these guidelines, and you'll move faster *because* you're steady, not in spite of it. That's the kind of professionalism that turns speed into trust.

Case study: How a firm used an AI-generated market scan to refine pursuit positioning

Two weeks before the proposal submittal deadline, the pursuit team at a 120-person design firm in the Pacific Northwest was stuck in a familiar place. They had a capable resume and strong relationship history, but their draft narrative for a major civic-center RFP read like every other competitor's:

- Design excellence
- Community focus
- Sustainability

True on all points, but nothing the client wouldn't expect from any competent team. In other words, the very definition of 'commodity.'

During a strategy session, one senior associate asked a simple question: *"What's this city actually talking about when they say community?"*

Instead of guessing, they ran a *ten-minute AI market scan*, asking for a targeted summary of public language pulled from the city's own news releases, budget statements, and planning documents from the past 18 months. The request was framed carefully:

"Summarize language used by <City Name> in public communications regarding facilities, parks, and community investment. Identify recurring phrases and priorities, and note any shifts in emphasis over the last two years."

The output was short but revealing. Across dozens of pages of civic announcements, the word *"belonging"* had replaced *"engagement."* References to *"shared economic corridors"* appeared repeatedly, which was language that pointed less to architecture as a building type and more to community as an interconnected system.

The team took that clue back to the pursuit room where their proposed theme shifted from *"a gathering space for the community"* to *"an engine for belonging."* Every piece of messaging including the graphics, callouts, and interview framing pivoted around that vocabulary.

When they walked into the shortlist interview, the selection panel chair told them, *"That word—belonging—that's exactly what we've been trying to capture."*

The project award wasn't just about the word itself. It was about resonance. The firm hadn't guessed what to say, they'd listened more efficiently. That's a huge part of the advantage of an AI-assisted market scan. It doesn't write the story for you, but it can show you which story your client is already trying to tell. By aligning with their current language instead of echoing last year's themes, the firm didn't just look prepared. They made it clear they understood.

That small story can help you reset the emotional temperature in your firm surrounding AI: Your professional credibility is maintained and even enhanced, not sacrificed, with this technology.

Try this now:
Create a one-page AI-assisted "Client scan sheet"

You don't need another template to manage. You need an easier and more comprehensive way to stay current with the clients who matter most. This exercise gives you exactly that: a single-page snapshot that blends human context with machine-gathered signals. It works best when you think of it not as a report, but as a *conversation starter.*

Step 1: Pick one client

Start small. Choose a client your team already knows well. Select someone you can reality-check quickly. The point isn't to find surprises, it's to practice seeing familiar ground through a clearer lens.

Step 2: Run a focused prompt

Use the same kind of prompt that kept our Marketing Coordinator Michelle efficient: *"Summarize the last 12 months of public news, press releases, and statements from <client name>. Highlight recurring priorities or phrases that reveal what success means to them."*

Add a line or two of context-related instruction such as: *"Focus on higher-education projects and community partnerships."*

In very short order—usually a matter of seconds—you'll have a digest of public signals.

Step 3: Pull it into human view

Open a blank document or shared slide. Divide the page into three columns and fill them with the insights that the report has provided. If you want, you can add these categories to the prompt and let AI sort the output into the three columns.

Signals	Patterns	Meaning
Short quotes, repeated words, metrics, title shifts.	Themes or trajectories you notice from the signals.	What those patterns could say about the client's evolving priorities.

You don't need to fill every box. One or two strong items per column is enough to see shape.

Step 4: dd your professional context

Now layer in what you *already know* from recent conversations, project feedback, and market chatter. Note anything that either supports or challenges the AI findings. This mix of machine aggregation plus human insight is the actual work of alignment. It's where raw data turns into valuable insight and perspective.

Step 5: hare it inside your team

Bring that one-page scan to your next internal meeting. Use it to open discussion with questions such as:

- "Does this sound like how the client's priorities have shifted?"
- "What implications does this have for our messaging or the themes we stress in our interview?"
- "Are there opportunities we've been missing?"

Notice the difference in tone. You're no longer talking about "what to Google"; you're talking about *what matters.*

Why it matters

A single page like this will save hours of browsing and countless dead ends. Its real value, however, is cultural. It teaches your team that valuable knowledge isn't about amassing sources or reams of data. It's about *distilling patterns together.*

Do one scan per week for a month and you'll start to see something bigger than data. You'll see a rhythm of understanding that keeps your work in tune with the market in near real time. When AI does the gathering and you extract the meaning, research stops being a chore and starts becoming a strategic muscle that is light, habitual, and human.

Summary: From insight to foresight

Chapter 5 has marked a turning point as we progress from *information gathering* to *insight-building*. We've seen how AI can surface the signals, while judgment, experience, and empathy create the meaning. Instead of speed, the goal has been clarity and the ability to see what matters sooner and speak the client's language more precisely.

Now that you can find and frame patterns, the next question is natural: *Can those patterns help you see what's coming?* That's where Chapter 6 begins.

We're going to learn about using AI to read the early signs that show up in the small public shifts, policy updates, and leadership changes, that hint at a client's next move. You won't need heavy analytics or expensive dashboards, just curiosity, light AI assistance, and the same interpretive confidence we've been building.

If Chapter 5 was about finding the story that's already in motion, Chapter 6 will be about sensing the *next* chapter before your competitors do, and turning that awareness into advantage.

CHAPTER 6

Predicting the client's next move

You read our minds

Following their shortlist interview for a major regional transit program, the pursuit team from a mid-sized firm on the West Coast was packing up their equipment and displays when one of the client panel members said quietly, *"It's like you read our minds."*

It wasn't a compliment about presentation style. It was recognition of alignment. The team's proposal and interview remarks had mirrored concerns the agency hadn't yet put in writing including funding phasing, equitable access, and the shift toward an electric fleet.

But the design firm's team hadn't guessed those priorities. They'd detected them.

Two weeks earlier, their marketing staff had used an AI-assisted scan to review recent city planning minutes, grant application updates, and public policy statements. A phrase surfaced repeatedly: *"transit equity corridors."* It wasn't mentioned anywhere in the RFP, but it appeared across three separate public updates by senior officials.

That single clue reshaped how the team framed its message. Instead of emphasizing general "community engagement," they described the design approach as advancing *mobility equity,* which was their way of solving both operational and social objectives at once.

When the panel used the phrase *"you read our minds,"* what they really meant was: *you saw the next question forming.* The team had met an emerging concern before it became a stated requirement.

That's the power of pattern awareness. It turns research into foresight and foresight into relevance.

In this chapter we'll explore how to build that capability intentionally and how to see beyond the stated RFP to interpret the early signals of what clients will need next. Because anticipation isn't intuition. It's empathy, backed by evidence.

Anticipation = Empathy + Evidence

Many marketers and business developers see anticipation as something that's almost instinctive, you either have or don't. In reality, it's a skill that grows from two repeatable behaviors: paying attention to people and paying attention to patterns.

Empathy without evidence is guesswork. Evidence without empathy is noise. Together, they create anticipation.

When you read a client well, what you're doing isn't magic; it's translation. You're connecting the spoken and the unspoken. The needs that appear in writing and the motivations that surface in tone, timing, or emphasis. That's why AI-assisted research, on its own, doesn't deliver foresight. It only gives you the fragments. Empathy assembles them in context.

Let's break the equation down.

Empathy: Seeing as they see

Empathy in business development isn't sentiment. It's perspective-taking. It goes without saying that you must see far beyond the technical scope of work and specifications, which is a rookie mistake that's common among technical professionals. To really connect with a client on a level that results in trust you must constantly pay attention to what matters in *their* world right now. These might include budget cycles, political pressure, leadership change, or mission drift. And then you have to understand how those conditions shape decision-making.

Empathy is also what keeps technology grounded. Machines can tell you what's trending, but only a person can infer *why* it matters this year.

Evidence: Listening beyond instinct

Evidence is the second half of the equation. It brings the discipline that balances intuition. It comes from building small, repeatable habits of inquiry including scanning public statements, noting language shifts, and tracking how priorities evolve across documents or interviews.

AI tools make this part much faster. They uncover the proofs that confirm or challenge your hunches. You still decide what deserves weight, but the model shortens the time between curiosity and confirmation.

The intersection: Calm confidence

When empathy and evidence meet, the result isn't prediction for prediction's sake. It's readiness. You start to feel the difference between reacting to requests and recognizing the direction before it's announced. You frame proposal ideas that make a client think, *"They understand what we're trying to solve,"* instead of *"They've seen our checklist."*

That equilibrium in which human understanding is refined by factual awareness, is where true anticipation lives.

Practical ways to anticipate change

Anticipating a client's next move isn't about building a data lab or hiring analysts. It starts with noticing small pieces of information such as budget notes, hiring updates, and shifts in tone that usually hide in plain sight. AI gives you the ability to see those pieces faster, so you can spend your time asking better questions.

Here are a few practical habits that turn that awareness into something you and your team can use right away.

1. Trend tracking: Stay current without getting lost

For years, staying informed about a client's market meant endless scrolling through trade journals and local news. Today, a simple AI summarizer can scan those same sources for you and return an organized digest in a few minutes.

Ask for something like: *"Summarize any new themes or recurring priorities in the last three months of news and public statements related to healthcare facilities owned and/or managed by <client name>."*

The goal isn't to collect every article. You're looking for movement and subtle changes in vocabulary or emphasis that show what's rising to the surface. When a government client shifts from saying "innovation" to "stability," or a university replaces "growth" with "belonging," those are early signals of how their priorities are evolving.

When you can capture and understand the shifts that are happening on a quarterly basis, you're keeping pace with both the client and their industry. That becomes your directional map and your advantage when it comes to your understanding of client priorities.

2. Relationship mapping: See how influence flows

In every client organization, information travels through people. When someone new joins a facilities board or a trusted consultant changes firms, the conversation around upcoming projects often shifts with them. You don't need any complex database to notice this.

Keep a simple chart of key contacts, their connections and their activities. What committees or boards are they serving on? What joint announcements have they made? What conferences are they attending?

If you ask an AI tool to *"summarize recent mentions of <Client Name> and key leadership names in public media or conference programs,"* you'll often see new networks forming before they're obvious in the RFP language. In fact, since clients are far worse at drafting useful RFPs than design firms are at responding to them, it's unlikely these important tidbits will ever find their way into any RFP.

You're not chasing gossip. You're orienting yourself and gaining vital insight into the individual, the client organization, and their industry. It helps you see where decisions are likely to be shaped and whose perspective may influence them.

3. Sentiment scanning: Listen for tone, not just topic

Clients announce priorities in many ways, but their *tone* often tells you more than their words. A city that suddenly sounds cautious in its press releases might be bracing for funding constraints. A university that starts using celebratory language again may have cleared a budget hurdle.

AI tools can scan public documents, social posts, or meeting minutes and give you a short tone summary. Are the messages positive, neutral, or critical? Taken over time, those summaries show "weather patterns" in public attitude, helping you approach the client with sensitivity instead of assumption.

4. Calendar triggers: Time your curiosity

Most agencies and institutions operate on predictable annual cycles of budget votes, grant deadlines, and leadership transitions. If you build, and then monitor a shared calendar around those events, you can then have an AI assistant run a quick pre-meeting scan:

"Provide any recent news or official statements from <Client Name> in the two weeks before their annual planning retreat."

That small habit gives you context before you reach out. You sound current, instead of reactive, and clients notice the difference.

Bringing it together

None of these tools are complicated. They're just a lighter, faster way to maintain the all-important context in which projects are initiated. So, when a client moves, you're not surprised, you're ready.

AI doesn't replace relational and situational intelligence, but it scales it with ease. With very little time and effort on your part, it keeps you close enough to the conversation that your next step feels natural, not forced.

Over time, these small patterns of attention build something much larger: a culture of awareness that makes every pursuit feel far more intentional and be far less subject to outdated or just plain wrong information.

From clues to conclusions to conversations

By now you've seen how small fragments of public information such as language shifts, personnel updates, and funding announcements can become early signals. The real advantage comes when you organize those signals into a rhythm your whole team can understand. That rhythm follows a simple, repeatable pattern: *clues, conclusions, conversations.*

Clues: What's showing up?

Clues are the raw discoveries. In other words, the things that your research or AI scans surface in the background. They might be a new phrase you've never heard from a client before, an unfamiliar partnership announcement, or a subtle change in tone.

The key discipline here is restraint. Don't rush to interpret every clue. Just capture it, label it, and keep it visible. When you start treating clues as data points instead of anecdotes, you begin to spot frequency and connection, which is the starting line for foresight.

Conclusions: What might this tell us?

Once a few clues start aligning, it's time to move to light interpretation. Ask, *"If these pieces belong to the same puzzle, what bigger picture are they hinting at?"*

At this stage you're testing hypotheses, not declaring truths. AI can help you cluster clues and surface possible explanations, but the judgment call to identify what's meaningful versus what's noise belongs to you and your team. It's the human conversation around the data that turns fragments into understanding. When you can transparently trace your reasoning from clues to conclusion, you gain confidence in your decisions and credibility in front of your colleagues.

Conversations: What do we do with it?

This is where the insight you're gaining can begin to pay off. Based on the clues you've collected and the conclusions they've led to, you can now bring those validated deductions into your internal team discussions and even your conversations with a client. Test them in planning meetings, proposals, and interviews.

Instead of saying *"we read that in the news,"* say *"we've noticed several indicators suggesting this direction. Does that align with what you're seeing?"*

Framing the question that way transforms static research into collaborative foresight. It signals your respect, attentiveness, and maturity. These are qualities that clients appreciate and remember long after a submittal.

Why the framework matters

Clues, conclusions, and conversations form a deceptively simple loop, but one that can change your firm's reflexes. Teams that follow these steps move from reporting facts to developing perspective. They start anticipating the next question instead of reacting to the last one.

The best firms have always done this, but the time and effort required to gather the clues, consolidate them, and reach reliable conclusions has been enormous. And, while naturally intuitive business developers have always had a nose for such insights, not everyone has those instincts.

AI can dramatically increase both the speed and the reach of this process. But it will always be your people who give it legitimacy. That combination of machine efficiency guided by professional judgment is what turns pattern awareness into partnership. And that partnership, more than any piece of software, is what wins work in this new era of practice.

Ethical sidebar:
Staying smart about privacy, scraping limits, and respecting human context

As you explore the use of AI to gather and interpret client information, it's worth stepping back for a moment. Foresight is valuable, but only when it's paired with respect for privacy, professional boundaries, and the human context behind every data point.

The biggest risk with these tools isn't technical; it's cultural. In a world where information is everywhere and automation makes it easy to collect, it's tempting to slide from *insightful* to *intrusive* without realizing it.

1. Stay on the right side of public information

Use what's publicly available including official reports, press releases, meeting minutes, social posts, and fiscal documents meant for open review. Avoid scraping private data or using work-around tools that claim to "unlock" restricted content. Besides the potential legal exposure, it erodes trust, which is the very thing business development is built on.

A simple rule of thumb: if the client intended the information for public eyes, you may analyze it. If not, leave it alone.

(I've included a sample prompt at the end of this section that you can use to ensure your searched remain on the right side of public information.)

2. Respect the spirit, not just the letter

Even within public data, context matters. For example, when scanning stakeholder comments from a community meeting, remember that every line represents a real person sharing a perspective. Use aggregated summaries ("common themes" or "recurring concerns") rather than quoting individuals verbatim, which can carry unintended sensitivity.

You can see an example of this on Amazon reviews. If you look at any product, you'll see an AI summary of the reviews. Instead of quoting an individual, it might say, *"Customers find this product easy to set up and use, although the quality receives mixed reviews, with some praising it while others find it marginal."*

Treat every data set as if your client were watching you work. That approach tends to produce decisions you'll never have to defend later.

3. Be transparent within your team

Make your research process visible. Document where information comes from, which AI tools you used, and how you validated what they delivered. Transparency builds internal trust and reduces the risk of misinterpretation. It also helps colleagues understand that these practices are about learning faster, not cutting corners.

4. Protect what you learn

Even when information starts as public, your interpretation may not be. Once your team draws conclusions about a client's potential direction, treat those insights as proprietary firm knowledge, not talking points to share casually. Professional integrity relies as much on discretion as on accuracy.

5. Keep human judgment at the center

No algorithm understands relationships, politics, or personal nuance the way you do. AI can organize and accelerate information, but it can't judge appropriateness. At the end of every scan or report, ask a simple question: *Would I be comfortable explaining how I got this information and why I used it?* If the answer is yes, you're still operating within the bounds of respect and professionalism.

The Payoff

The firms that will thrive in this next wave of business development aren't just the fastest. They're the ones their clients and their communities can trust.

Use AI tools as amplifiers of awareness, not shortcuts to advantage. Your credibility is your most enduring differentiator, and ethics are its proof.

How to prompt for responsible research

When you're asking an AI assistant to gather or summarize information, the way you phrase your prompt determines what it will (and will not) search. A clear ethical boundary, stated upfront, keeps the model inside professional lines.

Example prompt

"Collect and summarize information about <client or agency name> that is publicly available from verifiable sources only—such as official reports, press releases, meeting minutes, public websites, social-media posts made from official or public accounts, and fiscal or grant documents intended for open review. Do not access, infer from, or summarize any private, leaked, password-protected, paywalled, or otherwise restricted data. Identify the source of each item you reference."

Why this works

- It sets boundaries clearly: "publicly available" and "verifiable" are bright lines.
- It reinforces source transparency, which helps your team validate results.

- It trains the entire firm's AI use toward ethical consistency, not just compliance.

You can adapt this framework to any topic including projects, markets, competitors, and funding sources. The point is to embed responsibility into the process itself, so that ethical use becomes routine rather than reactive.

Putting it into practice: *The Quarterly Foresight Brief*

Once you start watching and listening for patterns, it helps to create a home for what you're learning. Otherwise, good insights end up scattered across emails and conversations, never shaping decisions.

One easy way to turn awareness into action is a *Quarterly Foresight Brief*. It's a short internal report of just two or three pages at most that gathers the signals you've noticed, the conclusions they suggest, and a few practical implications for upcoming pursuits. In a way, you can see this as supplemental information for your CRM.

You don't need to major in data science or hire new staff. All you're doing is capturing collective observation in a rhythm your team can sustain.

Step 1: Start with signals

Each quarter, assign one person on your marketing or BD team to gather inputs. They can use the tools we've already talked about including trend tracking, sentiment scans, and relationship mapping to surface clear, sourced snippets. Think of this as the *clues* section. There's no interpretation yet, just clean, verified facts.

A well-made brief might include:

- Two or three language shifts showing where client focus is moving.
- Notable leadership changes within key agencies.
- Emerging topics in recent public meetings.

- Community or stakeholder sentiments that could affect public perception.

  Keep it light. Half a page is often enough.

Step 2: Add context and conclusions

Next, invite a small internal group that might include marketers, project managers, and pursuit leads to review those clues together. Ask the two grounding questions:

1. *What might this mean for our clients this year?*
2. *How might it influence the kinds of projects or conversations we initiate?*

Capture those interpretations as a short *"So What"* summary. Bullet clarity beats paragraph density and the aim is to prompt awareness, not drown the reader in detail. We want to avoid the dreaded, TL;DR (Too Long; Didn't Read).

Step 3: Translate to action

Close the brief with two or three concrete actions or conversation starters. For example:

- Introduce the topic of mobility equity in upcoming transit pursuits.
- Revisit messaging on resilience to align with new state infrastructure grants.
- Plan a short check-in with City X before their budget hearings to understand how leadership changes might influence project timing.

  Now your brief moves from observation to opportunity.

Step 4: Share it broadly

Send the brief to all client-facing staff, not just marketing. Keep it in a shared folder. Treat it as an evolving playbook, not a polished publication. The best foresight documents don't read like reports; they feel like conversations your team was already having, just written down.

Step 5: Review and refine

After two or three cycles, patterns will begin to emerge. You'll start to see which signals consistently predicted change and which were irrelevant. That learning feedback strengthens your collective intuition while keeping it evidence-based. Over time, the *Quarterly Foresight Brief* will become part of your firm's rhythm as a subtle, easy mechanism that keeps everyone looking forward instead of just back.

Why this matters

When new AI tools appear every week, it's easy for teams to stall, waiting for the "next big capability." But maturity in practice rarely comes from adding technology; it comes from building habits that last. A small, thoughtful discipline like a *Foresight Brief* is a perfect example as you turn pattern recognition into predictable, teachable behavior.

That's how foresight becomes culture. It doesn't happen by shouting about innovation, but by normalizing awareness. Clients feel the difference in your questions, your proposals, and your timing. That's the kind of advantage no algorithm can automate.

Looking ahead: From prediction to presence

By now, you can see that predicting a client's next move isn't really about prediction at all. It's about presence in which you're close enough to a client's world that you can feel where their focus is drifting, and calm enough in your own outlook to respond with purpose instead of panic. That kind of presence, and the resulting insight, comes from solid habits, not access to some crystal ball.

The AI tools we've explored in this chapter are only accelerators. They help you see the same things your most instinctive business developers have always seen including context, intention, and early signals of change. But they make it possible for anyone across your firm to see them too. And to see them sooner.

When a firm develops that muscle, it stops thinking of business development as a chase for projects and starts seeing it as a continuous conversation that begins long before the RFP and extends well into the client's future vision.

That shift from transaction to anticipation is what steadies a firm in an uncertain market. It's also what sets the stage for what comes next.

In the next chapter, we're going to turn the focus inward. If Chapter 6 has been about *reading the client's signals*, Chapter 7 will be about *sending your own*. We'll explore how AI can help you listen to your firm's collective voice and the way your messaging, proposals, and people sound out in the world. Then we'll learn how to bring that voice into sharper alignment with the markets you want to lead.

Because being the firm that "reads minds" is only half the goal. The real power lies in becoming the firm whose own story is so clear, so consistent, that clients recognize themselves in it before you ever walk into the room.

Reflection: From foresight to presence

When you strip this chapter to its essence, anticipating your client's next move or the state of their collective mind isn't mysticism. It's discipline. You've learned that foresight lives at the intersection of empathy and evidence, and that AI's role is not to replace either, but to accelerate both.

Every strong pursuit begins when someone is paying closer attention than everyone else. Now, with simple tools and structured habits, that attentiveness scales from a single person's intuition to an entire team's rhythm of awareness.

Here are the key takeaways that you will want to carry forward:

1. **Pattern awareness beats guesswork**

Success doesn't start with bold predictions. It begins when you notice repetition. The best business developers aren't prophets. They're pattern watchers who translate subtle shifts in tone, timing, and vocabulary into relevance.

2. **Empathy is still the differentiator**
   While machines can tell you what changed, only you can understand why it matters. Empathy connects data to decision, turning disparate fragments into meaningful direction. Keep your curiosity focused outward on the problems clients are really trying to solve, not just what they've written down.

3. **Ethics preserve trust**
   The speed of new tools must never outrun the steadiness of your principles. Staying on the right side of public information, protecting confidentiality, and crediting human oversight aren't constraints. They're credibility multipliers.

4. **Habits outperform hunches**
   The *Quarterly Foresight Brief*, tone scans, and calendar triggers aren't novelties. They're good habits and even invaluable rituals. The firms that win tomorrow are those that make this awareness routine, not remarkable.

5. **Presence outlasts prediction**
   Anticipation isn't about outrunning your clients. It's about standing close enough to see where they're heading. That presence will earn trust in ways that no forecast can. It's the moment when that client says, quietly but memorably, "You read our minds."

Take a breath here. The prediction work is done for now. What comes next is projection as you learn to speak your firm's story with equal clarity and confidence.

In Chapter 7, we're going to turn from *interpreting* the market to *representing* your firm within it. You'll learn to refine your firm's collective voice so that empathy doesn't just flow outward, but resonates inward and back into the marketplace. Because the most credible signal you'll ever send is the one that sounds unmistakably like you.

CHAPTER 7

The voice of your firm: From insight to expression

It started with a document that no one wanted to read, least of all the people who wrote it.

A government client had requested an updated SF 330, and the team dutifully assembled theirs. It consisted of eight resumes, twelve sample projects, and fifty-two pages of polite monotony. When the lead principal flipped through the draft, she said, "It's accurate. But it doesn't sound like us."

Everyone nodded. Yet no one could quite define what *"us"* sounded like.

More than any edit or rewrite, that moment marked the turning point. The team realized that they weren't struggling with compliance; they were struggling with identity. The voice of their firm, diluted by years of transactional writing, had grown thin and generic. Their proposals now merely described *what* they did, not *who* was doing it.

The missing music

Every firm has a voice, even when it doesn't know it. It reveals itself in the rhythm of confidence, the cadence of empathy, the ratio of certainty to curiosity. Voice is how trust sounds on paper.

But voice rarely survives the endless iterations of bureaucracy. Over the years, layers of templates, disclaimers, and compliance checklists flatten tone until all that's left is the sound of facts covered with caution. The irony is that the closer teams try to mirror the "correct" format, the further they drift from the tone that made their clients fall in love with and then remember them in the first place.

This chapter is about recovering that tone. Not by becoming more creative, but by becoming more *consistent.* Because consistency *is* creativity when you're speaking for an enterprise instead of one person.

From intelligence to expression

The first six chapters of this book have built a foundation of awareness. We learned how to find accurate information quickly, interpret client signals responsibly, and anticipate what they'll need next.

Now all that foresight has to accurately and enthusiastically represent your firm in the outside world. It must take shape in your firm's visible materials including emails, qualifications packages, SF 330s, marketing decks, and proposal narratives. This is the moment when insight turns outward, and intelligence becomes *expression.*

You've already learned how to *see* like a strategist; now you'll learn how to *sound* like one.

Why is "voice" important now?

Before we can build better proposals, we have to improve how we write and communicate as a team. When a firm's proposals, interviews, and web copy all sound as if they've been written by different people, each with different personalities and viewpoints, over a long period of time, it signals disorganization long before any technical evaluation begins. Clients may not critique tone explicitly, but they *feel* the wobble.

Establishing the authentic voice of your firm first does three things:

1. *It focuses every line on trust.* Readers subconsciously hear continuity as competence.

2. *It reduces editing drag.* When tone is predictable, revisions move faster.
3. *It protects authenticity through compliance.* Especially in regulated contexts like the SF 330, where you can't design your way into distinction, sound becomes brand.

The public-sector game, in particular, rewards firms that can sound competent without sounding cold. Since you're not allowed to persuade overtly, persuasion has to live between the lines in clarity, rhythm, and respect.

Private-sector teams, by contrast, have space for narrative boldness. Yet they often lack the discipline that public work enforces. Voice bridges these opposites.

The chapter ahead

Here's what we'll do together in this chapter:

- *Find your baseline.* We'll explore practical AI-assisted scans to show you how your firm actually *sounds* today in its tone, word frequency, and sentiment balance.
- *Design a Voice Guide.* You'll build a one-page guide that sets guardrails for vocabulary, warmth, and rhythm.
- *Apply it under constraint.* We'll workshop how that same tone lives inside a government form, and how it sings in a private-sector design proposal.
- *Protect it in motion.* We'll finish with tools to keep that voice consistent as teams grow and projects multiply.

Because in the end, the goal isn't to sound inspired. The goal is to sound *dependable, familiar, and unmistakably you* even when the form itself seems intent on flattening personality.

Finding your firm's real voice

Every firm has two voices: the one it intends the world to hear, and the one it uses unintentionally, even accidentally. The first is how we *think* we sound, which might include confident, responsive, and even friendly. The second is what clients actually hear, which is made up of fragments that have been stitched together over time from dozens of contributors, deadlines, and templates.

If you've ever gone back to read one of your firm's proposals and felt a faint sense of deflation—something like, *"That's all true, but bland"*—you've already heard the gap between those two voices. The purpose of this section is to help you close it.

The voice equation

A firm's writing voice isn't built from one ingredient It's the result of three overlapping qualities: tone, consistency, and empathy. We could even express it as:

Voice = Tone × Consistency × Empathy

When all three reinforce one another, the firm sounds coherent and trustworthy. When any of them drop out, the voice loses power. And, while I've shown it as an equation, don't think of these factors as numbers, but as multipliers of believability.

1. Tone: The emotional temperature of a sentence

Tone is how your words *feel* when read aloud in someone else's mind. It's shaped by rhythm, verb choice, and sentence length. Most technical teams drift toward neutrality with long sentences, cautious verbs, and a tone of polite distance.

That's hardly surprising since technical professionals have been trained in technical writing, the primary purpose of which is to convey facts. The primary purpose of writing for marketing, business development, and sales, on the other hand, is persuasion. And that requires a different tone.

Clients rarely complain that a firm was *too clear* in a proposal, only that it was forgettable. Tone is the signal that first earns attention.

2. Consistency: The strength of the signal

Even a good tone will fail if it's inconsistent between documents. A website with a breezy, conversational tone can't be followed by a proposal that reads like a legal disclaimer. Clients sense that wobble long before they can diagnose it.

Where inconsistency produces "cognitive static," consistency tells your clients and readers, *"This is one team speaking with one mind."* It's how trust begins to form before anyone meets you.

3. Empathy: The direction of the signal

Empathy keeps your message pointed outward, toward your clients and the markets you want to serve. It means writing with awareness of what the reader must understand, not what the writer wants to display.

In public-sector contexts such as the SF 330, empathy appears not through emotion but through clarity and brevity with respect for reviewers who read hundreds of forms. In private proposals, it lives in framing and starts with *the client's challenge,* not *your firm's capability.*

- Tone captures attention
- Consistency builds credibility
- Empathy secures trust

When they align, your firm's writing develops a distinct, recognizable voice that is human, confident, and dependable.

Step 1: Listen to yourself

Before rewriting anything, listen objectively. Pull three samples of your firm's communication:

1. A recent proposal narrative (including an SF 330 if you have one).
2. A marketing brochure or thought-leadership piece.
3. A project description from your website.

Feed each into an AI language companion (ChatGPT, Claude, Gemini—whichever you use internally) and prompt it to: *"Analyze the tone, formality level, and emotional register of this writing. Describe what a reader might assume about our firm's personality based on this sample."*

Then ask the same tool to: *"Rewrite the preceding description in three words."*

You'll get summaries like *"cautious, thorough, distant"* or *"innovative, warm, concise."* Resist the impulse to argue. These are mirrors, not verdicts. Repeat for each sample and record the three-word outputs side by side. Patterns will emerge that are both flattering and uncomfortable. But those patterns are an accurate reflection of your firm's voice profile today.

Step 2: Compare intention to perception

Now gather a few colleagues, including business developers, project managers, a new hire, and anyone who writes on behalf of the firm. Ask them: *"If our firm were a person, how would it sound in conversation?"*

Capture adjectives, tone words, even body language descriptions. Then, compare those with the AI's summaries from Step 1. Where they match, you have alignment. Where they differ, you have opportunity.

You're not looking for poetic phrasing. You're looking for signal strength. Does your written tone transmit who you actually are? Or have layers of caution, iteration, and template compliance filtered out the humanity?

Step 3: Define the voice you intend

Condense that exploration into a short statement that will be the draft of your *Voice Guide.* It should be brief, with the absolute max being half a page. You can use this pattern to help you get started:

- *Our Voice Sounds:* Clear, steady, confident, and respectful.
- *We Avoid:* Vagueness, exaggeration, and unnecessary formality.
- *We Want Clients to Feel:* Understood, safe, and in good hands.

Keep it pragmatic. The fewer adjectives, the stronger the compass. You'll refine this guide in the next section, but even a rough draft starts to change how writers think. Suddenly, every email, project description, personal profile, or SF 330 sentence faces the same test: "*Does this sound like us?*"

Bridging both worlds

Public-sector writers often believe they have no room for tone. Private-sector writers often believe tone is everything. Both are half right.

The public sector enforces humility; the private one rewards personality. Voice is what lets a firm navigate between those realities without changing its character each time. It's how "accurate" becomes "credible," and how "friendly" becomes "trusted."

And it starts here as you hear your current sound clearly, without judgment.

Checkpoint reflection

As we wrap up this section and get ready to move to the next, here are some questions to prompt you and your team to reflect on this concept of *voice*:

- Which of the three variables, tone, consistency, and empathy, does your firm handle naturally?
- Where does it stumble?
- What part of your current written tone would you never want an AI to imitate?

Hold those answers as we move into the next stage where we'll translate intuition into shared direction so everyone, from Principals, to the Marketing Manager to the Proposal Coordinator, writes with the same voice.

***

Designing the Voice Guide

By now, your team has looked in the mirror. You've compared what your firm *intends* to sound like with how it actually comes across. The next step is to turn that awareness into something pragmatic: a Voice Guide, which is a short, shareable guide that keeps your tone consistent even when different people are writing under pressure.

You've no doubt heard of style guides or style manuals. These are standards for the writing, grammar, and formatting of written documents. In contrast, think of the Voice Guide as a style guide for trust.

What a Voice Guide is (and isn't)

A Voice Guide is not the same as a brand book. Brand books tell designers what colors to use. It's also not a writing manual instructing you about grammar conventions and spelling.

Instead, a Voice Guide is a one-page reference that explains *how the firm should sound* when it speaks in writing. It should be short enough for someone to glance at before writing an email or editing a proposal, but clear enough to anchor every external sentence your firm produces.

Purpose: To make the firm's voice transferable from principals to junior staff, from marketing to technical, and from public to private contexts without sounding mechanical.

Core components of a Voice Guide

You're going to structure yours around four simple sections and include concise, example-driven statements in each.

1. Tone

A short phrase describing the emotional quality you aim to project. Example: Calm, confident, and interested rather than insistent.

Add one or two "tone verbs" that illustrate how sentences should *feel to read.*

- We aim to sound like we're explaining, not selling.
- We avoid sounding like we're trying to convince instead of help.

2. Language guidelines

List a few cues on vocabulary and phrasing. Include both "lean toward" and "avoid" phrasing.

Example:

Lean Toward	**Avoid**
short verbs: *build, guide, deliver*	jargon clusters or compound nouns
simple, common language	Multi-syllable jargon or academic words
reader-centered phrasing: "You'll gain..."	self-centered phrasing: "We provide..."
active voice	passive voice

Keep this list to no more than five bullets each and always use the KISS principle.

3. Empathy markers

Identify concrete ways to show you understand the reader. Example:

- Begin paragraphs with the client's goal, not your firm's task.
- Use transitions that acknowledge constraints ("Because the budget was fixed…").
- Eliminate ego phrases ("our award-winning team") unless they are essential proof.

When writing for public-sector clients you can express empathy for the tough job of evaluating they face through order, clarity, and respect for the evaluator's time. Content for the private-sector can express it through narrative energy and shared ambition.

4. Signature soundbites

Capture a few repeatable phrases that sound exactly like your firm. These would be short sentences or tag-lines that can be used across various media.

For instance:

- Progress you can measure.
- Engineering calm into complexity.
- Design that builds consensus.

These become reusable "tonal anchors." When everyone uses them naturally, the firm's voice stabilizes without the need for constant policing.

Building the guide collaboratively

1. Gather a small team to create the first draft of the guide. This might include a marketing lead, one or two principals, and at least one project manager.
2. Start from what you already wrote in Step 3 ("Our voice sounds… We avoid… We want clients to feel…").
3. Refine language by consensus, not vote. Listen for the phrasing that makes heads nod. That's when you know you have resonance, not compromise.
4. Use AI as a calibration partner:
   - Paste the draft section into your AI tool.
   - Prompt it with: *"Rewrite this Voice Guide in the same meaning but in a slightly warmer/more confident/more neutral tone. Show three options."*
   - Discuss which output feels most "us." This keeps subjectivity grounded in examples rather than opinions.
5. Test before finalizing. Apply the draft to one live document such as a project sheet, a cover letter, or an SF 330 Section H paragraph. Ask the team: *Does this sound like who we said we are?*

Within one or two iterations you'll have language that feels natural, which is the test of a good voice.

Maintaining your voice over time

A Voice Guide will die quietly but quickly if it isn't visible. Once you've finalized it:

- Post it at the front of your proposal template or proposal guidelines manual
- Add it to the onboarding orientation for new writers and co-ordinators
- Revisit it annually or as your brand strategy or client base shifts
- Periodically re-run the AI tone check from *Step 1* to see if the tone of your written output still sounds like the one you decided you want.

Your firm's voice will evolve organically over time, so this exercise should be repeated occasionally. But if everyone is writing from the same guide it will resonate for your clients and your marketplace.

Example excerpt: A condensed Voice Guide

- ***Tone:*** Calm, confident, respectful
- ***Language:*** Active verbs, factual clarity; minimal jargon
- ***Empathy:*** Acknowledge the client's risk before describing our solution
- ***We avoid:*** Superlatives, clichés, and filler modifiers
- ***We want clients to feel:*** Reassured, understood, and informed.

That's all it needs to be. If your team reads that and immediately recognizes themselves, the guide is working. If the reaction is, *"we could never sound like that,"* you've just surfaced the real cultural work ahead. Which is exactly why we need to go through this process.

With a Voice Guide in place, you now have a tool to anchor every public and private expression of the firm's identity. Next, we'll bring that guide into the most demanding writing environment of all: the government's SF 330 form where voice has to exist inside the lines, but still come through as unmistakably human.

***

Sidebar: Humanizing the SF 330
Where warmth fits inside compliance.

The SF 330 is the great equalizer. Every firm writes into the same boxes, follows the same line spacing, and speaks through the same required sections. Which means the *words*, and not the layout, must carry all the differentiation. In these pages, the tone (or voice) you use becomes your brand.

When reviewers describe a submittal as "easy to read," they're really reacting to the humanity that you've embedded in the sentences, which ought to be clarity, rhythm, and consideration for their time. That sense of ease doesn't require casual informality. It requires careful empathy inside the format.

1. **Write for the reader's eyes, not the template**

The SF 330's boxes tempt writers to fill every inch. Don't. White space and shorter paragraphs create visual relief, especially in Section H. Aim for sentences that average 15 to 18 words. That's long enough to sound professional but short enough to retain rhythm.

Instead of:

Our multidisciplinary team collaborated cohesively to deliver a comprehensive design solution that met all programmatic and budgetary goals within the prescribed schedule.

Try:

Our architecture and engineering teams delivered a design that met every program and budget goal on time and without change orders.

Same facts, sharper cadence, less fatigue.

2. **Lead with the client's story**

Even within a compliance form, empathy starts at the sentence level. Frame your achievements as they relate to the client's challenge, not your firm's scope of work on the project.

Instead of:

Our firm provided civil and structural design for the new operations facility.

Try:

The city needed an operations facility that could handle growth without raising maintenance costs. Our civil and structural design made that possible.

This shift turns technical scope into evidence of listening.

3. Show coordination without using the word "coordination"

The word *coordination* appears hundreds of times across proposals. So often, in fact, that readers skip right over it. So, replace it with the outcomes that prove teamwork.

- *Weekly design sprints with city staff allowed early decisions on utilities and access.*
- *Co-location of disciplines reduced* RFIs *during construction.*

These statements show coordination without naming it.

4. Let AI help you hear the tone

Even in government forms, AI can act as a tone mirror. Try this safety-checked prompt:

"Rewrite this SF 330 Section H paragraph to improve clarity and flow, but maintain a formal, professional voice suitable for a federal submittal."

Then compare the AI's result with your original. Keep any phrasing that sounds more natural without becoming overly casual. Reviewers won't consciously identify the change, they'll simply keep reading.

5. Use your Voice Guide to check consistency

After editing, measure your text against the simple cues in your ***Voice Guide***:

Voice Guide Element	Quick SF 330 Check
Tone: Calm, confident	Do sentences sound factual without apology or hype?
Language: Active, plain	Does every paragraph move with verbs that act, not announce?
Empathy: Respect the reviewer's time	Could you read this aloud in one breath without tripping?

If the answer is "yes" to all three, the form is already human.

Remember: Clarity is warmth

Professional warmth doesn't come from adjectives. It comes from *time and energy saved for the reader.* Each clean paragraph, each declarative verb, and each respectful transition says, *"we care about how this feels to read."* That's the most reliable tone of trust you can send through a federal portal.

Bottom line:

You can't design your way into distinction in a format like the SF 330, but you can *sound* distinct through disciplined empathy. When your firm's submissions read with calm confidence and measured rhythm, reviewers will sense your competence even when you're not allowed to use logos, layout tricks, or color.

Lexical Density: When big words weigh you down

Engineers, architects, and technical professionals are trained to prove accuracy through vocabulary. It's a habit born from technical papers and specifications, but it sabotages readability when carried into proposals and client writing.

Linguists use the term *lexical density* to describe the weight of language. In other words, the ratio of long or complex words to plain ones. The more syllables, the denser and slower the reading.

The *Gunning Fog Index* is a tool that measures this density and returns a score that is equivalent to the number of years of schooling required to understand a piece of writing on first reading.

Publication	**Fog Index Score**
The Bible, Mark Twain, Harry Potter	~ 6
Time magazine *A Brief History of Time* by Stephen Hawking	~ 11
The majority of AE proposals	~ 18 - 20

That means your readers would need a PhD to find them easy to read. And here's the gentle truth: it's likely that you are among the guilty.

Technical professionals often lean toward words that sound more formal. Examples include *utilize* instead of *use*, *perform* instead of *do*, *inception* instead of *start*, *relocate* instead of *move*. While those choices feel precise, they rarely add meaning. They just add drag. Neil deGrasse Tyson and Stephen Hawking, both superb communicators of complex science concepts, remind us that the goal isn't to dumb down, it's to open up. The audience's time, not their intelligence, is the limiting factor.

When you trade grandiosity for grace, readers relax and trust what you say. And in proposals, trust wins out over terminology every time.

Chapter reflection: Sounding like trust

We began this chapter with the story of a lead principal who, having flipped through the draft of the SF330, said, "It's accurate. But it doesn't sound like us." She wasn't talking about commas or fonts. She was hearing the erosion of confidence that happens when a firm's writing loses its rhythm and its voice.

Recovering that voice is what this chapter has been about. Trust, after all, is rarely built through new vocabulary; it's built through *familiar consistency*. And it shows up in the same calm, respectful tone every time your firm speaks, no matter the medium or the deadline.

Sounding like trust doesn't mean sounding "friendly." It means that your words carry a steady pulse of competence without friction or strain. It means clients don't have to work to understand you. And when language feels effortless to read, people assume the thinking behind it is solid, which, most likely, it is. It's the reflection of a firm whose competence, knowledge, and insight are so solid, they can afford to talk about their work in the relaxed, everyday language with which clients are comfortable.

In an age of AI-generated sameness and templated replies, a firm that writes with human rhythm immediately stands apart. There's no trick to it. It's simply tone, consistency, and empathy, repeated until they become reflex. When your voice is clear enough that a reviewer leans into it rather than away from it, you've achieved the best form of persuasion: credibility that doesn't have to advertise itself.

Next, in Chapter 8, we'll turn from *how* your firm sounds to *how* your communication is structured with the goal of transforming that clarity into proposals that persuade through order as well as tone.

CHAPTER 8
Proposal intelligence: Structure that sells

When your firm's voice has found its rhythm, the structure of your proposals has to keep the beat. Your firm's tone may be confident, clear, and empathetic, but in most pursuits, *structure* is what actually sells. Order, logic, and flow shape the reader's experience long before your words have a chance to persuade.

Most proposal teams approach structure as a compliance exercise: the RFP lists a sequence of requirements, and writers respond to those instructions one section at a time. Technically, nothing is wrong with that, except that it leaves no room for *coherence.* Readers wind through fragments of information instead of following a story.

Our goal in this chapter is to restore coherence and connection without ever violating the rules. Because proposal intelligence isn't about writing faster. It's about thinking structurally, and building a framework for the content of your proposal that carries meaning, momentum, and the human signal of trust.

From compliance to coherence to connection

First, let's define the terrain.

1. **Compliance**

This is the minimum threshold. While it keeps you eligible, it's not a guarantee that you'll be competitive. Compliance ensures your content matches the RFP outline, answers every instruction, and respects word limits and forms.

AI can excel here as it assembles resumes, formats tables, and cross-checks data against requirements, confirming that all the "have you included..." boxes are ticked. But compliance doesn't persuade. It only opens the door to persuasion.

2. Coherence

Coherence creates a logical flow. It arranges all the compliant parts so they make sense together. It turns forty independent paragraphs into a single narrative arc. Coherence asks: *Does one idea lead naturally to the next? Can a reviewer follow our thinking without backtracking or guessing?*

This is where the people lead and AI assists, helping you test sequence, identify repetition, and check transitions.

3. Connection

Connection is the top tier. It's the emotional logic that makes your story memorable. It's where human perspective transforms information into relevance.

Connection answers the reader's unspoken question: *Why should I trust you with this problem?* It lives in the tone of the writing, the examples you cite, and the framing of your narrative. It's the creative proof that separates a compliant submittal from a compelling one.

While AI can't invent connection, it can reveal patterns that point you toward it with the repetition of key themes, unusual word frequencies, or narrative gaps where empathy could live.

The chapter ahead

In the following sections, you'll see this framework applied across both public and private contexts:

- *AI's role:* assembling data, resumes, and project sheets; detecting inconsistencies; auditing submissions for alignment.
- *Your role:* story architecture and emotional logic using the structure that makes facts persuasive.

- *Public Sector Deep Dive:* mastering the SF 330's unique constraints, including section-by-section guidance and "creativity within the box."
- *Private Sector Parallel:* storyboarding, visual balance, and freedom of voice without losing focus.
- *Ethical Sidebar:* disclosure and diligence when using AI on regulated submissions.
- *Mini Case Study:* how one firm used AI to unify tone and structure across 80 projects.

Then we'll close with some thoughts about precision, the all-important technique that makes complex material effortless to read. Just as Chapter 7 showed that *trust sounds human,* this chapter will show that *trust also looks organized.*

AI's role and yours

AI is a brilliant assistant but a terrible author. Its strength isn't imagination, it's integration as it brings together scattered information so you can think clearly and act decisively. Instead of asking AI to "write the proposal," your best strategy is to treat it as a *chief of staff for information.* It's brilliant at tracking details, testing alignment, and flagging weak spots before your reviewers do.

What AI does best

1. **Gathering and sorting data**

AI can locate and compile resumes, project sheets, and boilerplate descriptions in seconds. A simple prompt such as *"List all project sheets mentioning LEED Gold certification completed after 2020"* can replace hours of manual searching.

2. **Detecting inconsistencies**

Ask your system to audit resumes, project lists, and narratives for mismatched dates, project titles, or terminology. *"Compare Project X across all sections and report inconsistencies in scope, completion dates, or job titles."* This isn't creative work. It's quality assurance, and AI can perform it relentlessly and accurately.

3. **Converting data into structure**

   Once content is gathered, AI can outline or cluster it according to the RFP sections or evaluation criteria. Prompts such as *"Group these project examples under the evaluation headings in Section M"* save early drafting time and reveal content gaps.

4. **Tone and redundancy checks**

   While AI shouldn't compose your narrative, it can mirror tone and flag excess. *"Identify sentences longer than 25 words"* or *"Highlight repetitive phrases related to teamwork."*

   You can use mechanical filters like these to keep the human writer's focus sharp.

Where you – not AI – must lead

Your people must still carry the irreplaceable responsibilities for story, logic, and persuasion.

1. **Story**

   You decide which projects define the pattern, which details deserve space, and which anecdotes support the client's outcome. No algorithm can sense the moment when your firm's experience "just feels right" for this project.

2. **Emotional logic**

   Data proves competence, but emotion proves confidence. The phrasing you use shows how you understand the client's risk, urgency, and hope. That resonance, which you deliver through rhythm, tone, and framing, belongs exclusively to the flesh-and-blood writers who've lived through the pressure and the pride of the project.

3. **Ethical and strategic judgment**

Machines can map rules and instructions, but only people can weigh when bending them improves readability or when compliance is absolute. The more AI evolves, the more valuable human *editorial conscience* will become.

In practice

You can envision the collaboration between AI and your team as a Venn diagram.

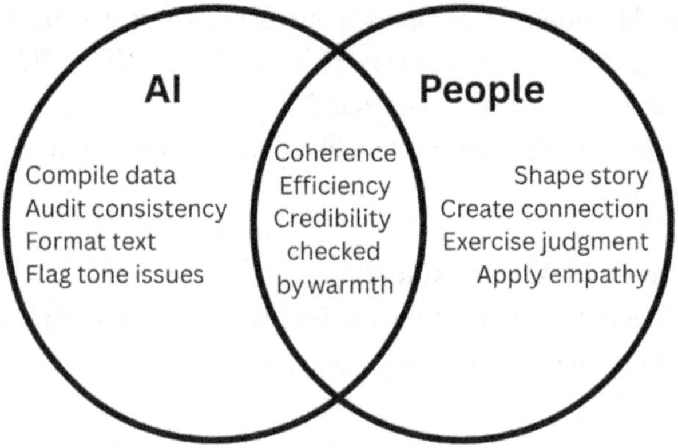

When the strengths of the two are combined, they keep your proposals both compliant *and* coherent, so you can devote your limited time to the work only a person can do, which is forging connection.

Now that you know what each partner contributes, let's take a deep dive in the public-sector where compliance rules are strict, templates are unforgiving, and "creativity in a box" becomes the defining skill.

Public Sector deep dive: Mastery of the SF 330

The SF 330 is a paradox: it rewards innovation yet punishes deviation. Everything about its design is intended to make firms indistinguishable. The fixed boxes, standardized section titles, and uniform fonts line up all the firms in a quickly scanned row.

And yet, every successful submission has its own internal structure. That's where proposal intelligence can show its power by using AI for discipline and people for design—not design in the visual sense, but design in the sense of *structuring thought.*

1. **Use AI as your compliance controller**

   It's almost as if AI were built for the SF 330's mechanical precision.

   - **Checklists to guard against human drift**

   Feed the solicitation into your AI tool and prompt: *"Extract and number every sub-requirement from Section E through H."* The resulting list becomes your audit map.

   - **Document cross-validation**

   Prompt it by asking: *"Compare the project names and dates in all resumes with those in Section F project descriptions. Flag discrepancies."* What once took an hour of line-by-line comparison can now be done in seconds.

   - **Consistency in formatting and phrasing.**

   Use controlled prompts: *"Make all role titles in Section E parallel in wording and capitalization."* This kind of invisible discipline is what allows reviewers to read without the "friction" that always slows and distracts comprehension.

2. **Use people as story builders**

   Once you've assured compliance with the rules and requirements, the real craft begins.

   - **Shape the logic, not the layout**

     Every SF 330 forces you to follow the same framework. Your advantage lies in the *sequence of emphasis.* Lead each project story with the *client problem*, not the design solution. Empathy that's built into the SF330's structure will always outshine structure alone.

   - **Re-order the reader's experience through transitions.**

     While you can't change section order, you can

choreograph momentum. Phrases like *"That same approach informed…"* or *"Building on this success…"* connect boxed information into an implied storyline.

- **Embed rhythm.**
  Vary your sentence length, but stay concise. Each line should move with purpose and breathe easily when read aloud. Mix short, factual statements with a few slightly longer ones that connect ideas; that ebb and flow creates the rhythm that keeps readers engaged. Reviewers often read aloud quietly to themselves. Your job is to meet them with writing that sounds natural, not endless.

3. The AI audit loop

Treat your SF 330 process as a living feedback system:

a) *Create the first draft manually.* Capture your firm's authentic voice, rhythm, and clarity.

b) *Run through an AI audit.* Ask for readability scores, redundant phrasing, or overly long sentences.

c) *Human-edit for warmth.* Restore your firm's tone where the machine flattens flair or gets too flamboyant.

d) *AI-check again.* Confirm compliance, character counts, and alignment to instructions.

e) *Final human review.* Read aloud once; anything that sounds wrong will surface.

This loop ensures every proposal passes the "two intelligences" test before it leaves your desk.

4. Where creativity actually lives

Even inside a federal form, creativity can survive and amplify your message. It just does that differently than in other documents.

- *In sequencing:* choosing which projects are best to tell the client's story.

- *In compression:* distilling your experience and achievements into active, elegant phrasing.
- *In empathy:* designing the reading experience for people operating under pressure and deadlines, not just for the contracting file.
- *In consistency:* using repeated structural motifs that make your form easy to follow.

When every competitor is forced to use the same form, the humanity you bring to it *is* your creativity.

Public sector proposals have a well-earned reputation for constraint and precision. But the same discipline can make your private sector proposals more persuasive. In the next section, we'll shift to the private-sector, where storyboards, visuals, and a freer tone can amplify the very structure you've mastered here.

***

The Private Sector:
Storyboards, visuals, and freedom of form

While it may appear that proposals for the private sector allow you much more freedom than government forms, the absence of strict instructions can lead you to wander. Without limits, teams often substitute volume for clarity.

Proposal intelligence means channeling your creative freedom with the same discipline you honed inside the boxes of the SF330. But now you can express yourself in color, hierarchy, and pacing instead of just numbered forms.

1. **Start with a storyboard, not a template or a 'Save-As' of a previous proposal**

Before opening InDesign, PowerPoint, or a previous proposal that you intend to copy, sketch the *arc of the story you want to tell.*

You can use AI as a rapid-draft partner by prompting, *"Summarize the client's stated priorities and list three logical narrative orders for a proposal responding to them."*

When you review the three options, you'll instantly see which one best mirrors how the client will think while reading your proposal. Then, *human-edit the path* by asking where do you want the reader to be curious? Where will you need to provide evidence? Where will you need to offer assurance? That's how you will build your storyboard.

A practical storyboard outline might follow a flow such as this:

a) *Challenge:* Define the client's pain, need, or opportunity.
b) *Insight:* Why it matters and what's at stake.
c) *Approach:* Your team's guiding principle or process.
d) *Proof:* Evidence, including projects, metrics, and credentials.
e) *Commitment:* Partnership tone, next steps, confidence.

Every slide or paragraph should support one of those beats. Anything that doesn't is extraneous noise and will distract from your message.

2. Use AI for shape but let the people give it meaning

While AI can produce outlines, draft transitions, and even generate image placeholders when a concept calls for a graphic, only people can decide the intent of your sequence. That's the "why" behind the order you've chosen.

When an AI tool suggests, *"Combine these two sections — both discuss project management,"* you should always step back and ask: *Would collapsing them strengthen our story or flatten it?*

Proposal intelligence isn't just accepting the suggestions AI provides, but questioning them against what you're trying to accomplish.

3. The design of your proposal is structure you can see

In private proposals, the layout of the document becomes a *visual grammar* that invites the client to trust you. That means you should use design cues not as decoration but as wayfinding.

- Headings = movement. Keep levels consistent so a reader's eye can travel through your document effortlessly.
- White space = vital breathing room. When every block screams for attention, nothing speaks clearly.
- Visuals = clarity. Graphs, diagrams, and photos should replace words, not repeat them. AI can help re-balance pages by calculating text-to-image density or suggesting optimal word counts per section. Then the people can decide whether that density still "feels" readable.

4. Tone and tempo

Your voice and the structure of your proposal document must stay synchronized and there are some simple techniques you can use to accomplish that. For example:

- Use short, declarative summaries to anchor each major section, especially after long visuals.
- When switching topics, add brief re-orientation lines ("Here's what that means for you…") to preserve coherence across visual breaks.

Think of the tone of the document as the audio soundtrack behind your visual layout: if it's erratic, the photos and graphics won't save it.

5. Audit the document before releasing it

Before finalizing your proposal use the same dual-intelligence loop you refined for working with the SF 330:

- *AI Review:* analyze reading time, identify pages overly heavy with text, flag repeated phrasing.
- *Human Review:* re-introduce warmth and pacing, make sure every headline accurately reflects its section.

This is still proposal craftsmanship—only the tools have changed.

Preparing a proposal for public sector work will teach you discipline while doing the same for a private sector project can reward your imagination. When those two skills operate together, as discipline shapes imagination, you move from a proposal structure that simply conveys facts to one that *persuades.*

Next, we'll pause to consider the *ethics* of this People/AI partnership: how to disclose, document, and manage AI's role responsibly in pursuit materials that represent licensed professionals.

Ethical Sidebar: Disclosure and diligence

As AI tools accelerate proposal creation, your firm must match that speed with ethical transparency. In regulated professions such as architecture and engineering, the integrity of your submission holds the same weight as the drawings and calculations that follow it. These aren't optional niceties. They're the framework that keeps innovation credible.

Proposal intelligence is more than efficiency. It's also stewardship of truth.

1. Disclose AI use where it matters

If AI contributes to any portion of your submission, especially in narrative drafting, resume assembly, or past performance summaries, acknowledge that assistance in your project records or an internal proposal log. When a client or agency requests disclosure, adapt that record into a brief statement within the proposal itself.

Public agencies increasingly ask whether AI has been used in preparation of materials. Even where it's not specifically required, documenting your process protects both the firm and the licensed professionals who sign the submittal.

A simple internal note such as, *"Sections E and H quality-checked through generative language audit on <platform name>, reviewed and edited by <person name>"* is enough to demonstrate accountability and human oversight.

Remember: disclosure builds confidence. Secrecy invites suspicion.

2. Preserve data diligence and client confidentiality

Never upload proprietary or client-sensitive information into a public AI model. The duty of confidentiality applies to all content, including narrative drafts, project databases, and résumé details.

Before using any AI tool, confirm:

- Where the data is stored and how long it's retained.
- Whether the provider uses your inputs to train future models.
- That your firm's confidentiality agreements cover third-party digital tools.

If you want to learn how to confirm where your data goes and how long it's kept, I've provided instructions further along in this section.

If any answer is uncertain, default to offline, locally managed tools or anonymized datasets for prompts. The absence of a non-disclosure clause from an AI vendor is your signal to treat the platform as a public forum.

3. Maintain the human seal of responsibility

When a licensed architect or engineer signs a proposal, he or she is endorsing its accuracy and integrity, not just its design approach. AI can assist in finding inconsistencies, but it cannot *certify* compliance or professional judgment. That authority remains, and must always remain, with the professionals on your team. Use AI for its analytical eyesight, not as a signer of record.

Before you submit, always ask: *"Could I defend each line as true and fair if questioned under oath?"* That's the ethical test. AI won't be in the hearing room with you.

4. Audit for bias and representational fairness

Generative systems learn from large data pools that may over- or under-represent certain project types, communities, or disciplines. Before publishing an AI-drafted summary, read for balance:

- Do we reinforce unintended stereotypes?
- Does the description credit all team members accurately?
- Are we overstating the firm's role in past work?

Fair portrayal isn't cosmetic, it's character in written form.

5. Document the boundaries

Adopt one firm-wide statement on how AI will and won't be used in pursuit documents. Clarity upfront prevents discomfort later. For example:

Our firm uses artificial intelligence tools to assist with data compilation, proofreading, and consistency checks. All narrative sections are reviewed, edited, and approved by licensed professionals before submission.

That single paragraph, kept on file or in boilerplate disclaimers, signals maturity and foresight to reviewers, clients, and staff alike.

How to confirm where your data goes and how long it's kept

Note: The following guidelines are practical checks for professional diligence, not legal advice. Always verify requirements with your firm's counsel or compliance officer.

1. **Start with the platform's published policy.**

- Every reputable AI vendor maintains a *Privacy Policy* and *Terms of Service*. They are usually linked in the footer of their website or inside the user-account area.
- Search the page for keywords like *"data retention," "training," "storage,"* or *"third party."*
- Look for plain statements such as "we may use your inputs to improve our models" or "user inputs are not retained after session termination."

2. **Check for an enterprise or business version.**
- Many AI providers offer a "business," "enterprise," or "pro" plan where inputs are kept private and not used for training.
- Your IT or compliance lead should request the vendor's *data-use statement* or *SOC 2/ISO 27001 certification* if available.
- If no enterprise documentation exists, treat the platform as public space.

3. **Ask directly.**
- If policies are ambiguous, send a short, written inquiry to the vendor's support or sales contact. Your inquiry might read: *"Our firm handles confidential project data. Could you please confirm whether prompts and uploaded content are used to train your models or otherwise stored after a session?"*
- Keep the response with your internal records of AI tool vetting.

How to ensure your firm's confidentiality agreements cover third-party tools

1. **Review your existing NDAs and client contracts**
- Many template NDAs restrict you from sharing *any client information with unaffiliated parties.*
- Uploading that content into an external AI system counts as "sharing" unless specifically allowed.

2. **Consult your internal counsel or risk manager**
- Ask them to insert a clause in your firmwide technology policy specifying that **AI tools may only be used on content cleared for public or internal distribution.**
- An example clause might read: *"Team members shall not input client-confidential data into any AI tool unless the platform's terms of service expressly prohibit reuse of inputs and data storage beyond session purpose."*

3. **Maintain a short "approved tools" list**
- IT or your legal counsel can vet AI platforms, record data-security assurances, and post an approved list for proposal staff.

- This prevents teams from having to run the same risk analysis individually.

Ethics is not a pause button on progress. It's the governor that keeps trust from burning out at the speed of technology. Every one of your clients is aware of the inroads that AI is making in the workplace. They're likely working to integrate it themselves and won't be surprised to learn that you are too. But the more transparent your methods, the more your clients will trust the results.

With these ethical guardrails in place, we can look outward again and see how your careful attention to the way you structure your proposals creates unity across your firm's collective output. In the next section, we'll explore a brief case study: how one team used AI to unify tone and structure across 80 projects without losing the human voice that made them distinct.

***

Case study: Coherence at scale

A mid-sized, multi-discipline firm in the Southwest wanted to bring eighty past project descriptions into alignment for an upcoming pursuit. Each description had been written by a different project manager over several years. Some were too informal, some stiff, and many were redundant. Volume was only part of the challenge. The bigger obstacle was consistency of *voice.*

1. **The problem**

The firm's marketing team realized that while every project told a credible story on its own, the collective portfolio read like eighty different firms. Terminology, formatting, and even tense varied from page to page. Reviewers had to keep re-orienting themselves, which made their otherwise strong experience feel disjointed.

2. The approach

They built a simple workflow using the same "two-intelligence" model introduced earlier in this chapter.

- **The AI pass for standardization**

The team uploaded all project narratives into a secure, enterprise AI workspace. They prompted it to: *"Identify format inconsistencies, repeated phrases, and divergent narrative structures across these eighty project descriptions."*

The system produced a uniform checklist of recurring issues including verb tense, project-size units, and inconsistency in "client" vs. "owner" language.

- **The human pass for tone and emphasis.**

With the AI audit complete, senior marketing staff reviewed the output to decide which patterns improved clarity and which diluted personality. They then wrote a *tone brief* describing how the firm should sound: *confident, relational, succinct.* The brief became the reference document for all revisions.

3. The execution

Each project lead revised two or three records using the AI suggestions and the tone brief as dual guides. The system flagged long sentences, aligned subheadings, and ensured metrics and awards appeared in consistent order. Humans verified the transitions and inserted empathy cues using phrases that tied outcomes back to client impact.

To prevent drift, they ran a final automated audit prompting the AI assistant to: *"Find variations in section sequence, inconsistent terminology, or missing key data fields."*

The entire correction process was completed in hours instead of weeks or months.

Actually, if we're going to be completely truthful, "weeks or months" is far more likely to mean "never." Most firms would never attempt such a revision, knowing that, using conventional methods, the time and effort involved would be simply overwhelming. The introduction of AI assistance brings those "some day…" to-do items into the realm of "entirely possible."

4. The result

Within one week, all eighty project pages shared a recognizable rhythm. When any group of them were included in a proposal, reviewers could sense a unified voice that was clear, courteous, and credible without the monotony of machine uniformity.

The marketing team later integrated the same tone brief into the firm's website and interview prep materials; *proposal intelligence* had broadened to become *firm intelligence.*

"We didn't lose our humanity," the marketing director noted, *"we just gave it a better filing system."*

Consistency doesn't flatten individuality. It focuses it. AI assistance gave the firm both speed and discipline. The people gave it perspective and purpose. That partnership produced the most persuasive approach of all, one that *sounds* organized and *feels* human.

***

Precision: The final layer of trust

A great deal of the "art of persuasion" results from what we might be tempted to call *intangibles.* As I've repeated often so far in this book, and in the many books I've written previously, every one of your competitors is fully capable of executing the project you're currently chasing. They've got the qualifications, experience, and judgment to do the job. So, your clients aren't just looking for competence, they're looking for that something extra that will put your firm ahead of the others and make their decision easy.

One of the most important of those intangibles is trust. And trust doesn't just *sound* human. It also *looks* organized.

Precision is the quality of your proposal that makes its complex content read simply and easily. It's a behind-the-scenes organization that turns complicated and multi-faceted information into fluent understanding. When coherence and connection work together, the precision of your proposal holds them steady.

1. **Precision is clarity under pressure**

Proposal deadlines are rarely generous. Teams hurry, copy, paste, and merge sections, and promise to "clean it up later." Precision is what saves your message from that empty promise.

It's not about correctness for its own sake; it's about the readability of a document that's been assembled under stress. A precise proposal is one your reviewer can skim at 11 p.m. and still understand perfectly.

Think of it as the *civil engineering* of persuasion. It's the load-bearing language that keeps the story upright no matter how many revisions it endures.

2. **Where AI helps**

AI excels at the mechanics of precision:

- Spotting inconsistencies in terminology and units.
- Flagging sentences that exceed readability thresholds.
- Counting passive-voice constructions or filler phrases.
- Checking for "orphaned" references, missing attachments, or mis-numbered tables.

That's not just polish, it's structural integrity. When a tool can handle the micro-corrections, people stay free to focus on meaning.

3. **Where the people need to lead**

Precision at the human level is judgment in choosing which details earn space and which distract. While a machine can eliminate error, only a person can recognize elegance, the moment a sentence feels just right, or an idea is expressed in exactly the way the reader will understand it on first try.

Ask these questions before hitting Send:

- Does every line in this document have, and perform its job?

- Could a tired reviewer find answers without re-reading?
- Does the rhythm feel predictable but alive?

If the answer is yes, precision is doing its work.

4. Making precision a habit

To build precision into every one of your proposals, no matter how rushed the process may be, create a lightweight, repeatable checklist for every proposal.

a) *Language pass:* Precision in tone and terminology.
b) *Logic pass:* Precision in flow; check cause-and-effect sequence.
c) *Format pass:* Precision in visuals and spacing; does the eye rest where you intend?
d) *Empathy pass:* Precision in relevance; does every section connect back to the client's needs?

The first three can and should be performed in seconds by AI. The fourth step, performed only by a person, is the one most firms skip. It's also the one that will resonate most with clients.

The payoff

When proposals are clear, consistent, and grounded in empathy, clients don't just *read* them, they *trust* them. The precision of your document transforms its organization and layout from an outline into a signal: *"If they write with this much care, they'll manage our project with the same discipline."*

That impression is the real sale.

Closing reflection

Proposal intelligence isn't about replacing people with process or technology. It's about making use of the input of *both* in service of clarity, coherence, and trust. Compliance keeps you eligible. Coherence keeps you readable. Connection keeps you memorable. And Precision (the quietest virtue of all) keeps you believable.

In the chapters ahead, we'll see how these same principles scale beyond proposals into your firm's broader communications, including reports, interviews, and the stories that define your practice in the marketplace.

PART 3

WINNING THE MOMENT: PROPOSALS, PITCHES & PERFORMANCE

CHAPTER 9
The modern proposal formula

Re-framing the formula

The way we build proposals hasn't kept up with the way clients read them. Most firms still use layouts that were designed for print and PDF: they're tidy, linear, and heavy on compliance. That approach worked when reviewers flipped through pages on their desks. Now they scroll, search, and scan. What used to look organized now just feels old and slow.

We need something quicker and more flexible: a *framework* that adapts each time, rather than a template that freezes the process in place.

AI hasn't only changed how proposals are written. It's changed how they're judged. Reviewers, software, and search tools all look for patterns that signal clarity and consistency. If your proposal doesn't read well on screen or make sense to both a person and a system, you've already lost ground.

A modern proposal must do three things at once:

1. *Be easy for software to read.* Organized, searchable, tagged, and complete.
2. *Be easy for people to read.* Clear, direct, and logically connected.
3. *Build trust.* Through tone, empathy, and attention to what really matters.

Those three layers, technical, structural, and emotional, work together like the foundation, framing, and finish of a good building.

Why templates fall short

Templates were built to save time, not to tell stories. They keep things consistent but often flatten the content along the way. Over time, they turn unique pursuits into look-alike packages.

If you drop an outdated template into an AI tool, the tool can repeat the same problems even faster. The output will be perfectly consistent, but equally forgettable. The result is more repetition, not better persuasion.

Firms that stand out today don't rely on static templates. They rely on *awareness of pattern.* They emphasize the ability to see how information connects and flows across different pursuits, not just how it fits into boxes.

From templates to frameworks

A framework teaches you *how to think* about structure, not just *where to paste* your content.

- A template divides space and locks you into a sequence.
- A framework defines purpose and helps you decide the sequence that fits the story.

That shift, from filling blanks to designing flow, is what makes a proposal both compliant *and* compelling.

A better starting question

Before you start any pursuit, I encourage you to skip past, *"Which template should we use?"* and instead ask: *"What path will lead this client from problem to trust?"*

That one question changes everything. It puts the story you need to tell ahead of formatting, and empathy for your client ahead of efficiency in proposal prep. That's how a modern proposal earns attention in a crowded field of good intentions.

Next, we'll take this formula apart, layer by layer, machine-readable, human-readable, and emotionally persuasive, to show how AI and your team can work together inside each one.

The three layers of a modern proposal

I already said this at the beginning of this chapter, but it deserves repeating: Every competitive proposal today has three jobs to do:

1. Make sense to software
2. Make sense to people
3. Make people *trust* what they're reading.

Those layers overlap, but each one deserves its own attention. Think of them as the three dimensions of readability; technical, visual, and emotional. When all three are balanced, your proposal doesn't just check boxes, it actually connects.

1. Machine-readable: How software "sees" you

Most reviewers still read proposals themselves, but before that happens, many submittals go through an early digital pass. Some systems scan for keywords that match evaluation criteria. Others extract and organize data so reviewers can compare firms side by side. Either way, you're often being read by a machine before you're read by a person. That means your content needs a little structure under the surface:

- Use consistent headers and section names that match the client's language.
- Put important facts in regular body text as well as in charts or graphics. Anything that's only shown as part of an image may not be searchable or picked up by your client's software.
- Tag or label resumes, project sheets, and attachments in ways that mirror the RFP's sections.
- Keep dates, roles, and metrics in a clean, recognizable format.

You don't have to write like a robot. You just need to make sure a robot can *follow the trail.* Once that baseline is solid, your human readers can focus on what matters most.

2. Human-readable: How people follow your story

People don't read proposals like novels. Instead, they jump, skim, and sample. Every sentence is a crossroads—either the reader leans in, or they skip and move on. Coherence is what keeps them leaning in.

A few ways to make it easy:

- Open each section with a short orienting line that says, in plain terms, what this part covers.
- Keep paragraphs short and focused on a single idea.
- Use subheads that sound like promises, not just labels. Try, *"Results That Last"* rather than *"Outcomes."*
- Repeat key phrases from the client's stated priorities so reviewers see alignment at a glance.

When you do this well, reviewers can track your logic without extra effort. They start to trust that your team thinks the way they think. And that trust builds steadily, paragraph by paragraph, and section by section.

3. Emotionally persuasive: Why they should care

Once the facts and flow are in place, the emotional layer gives everything its lift. Your readers aren't only measuring technical skill; they're measuring *fit*. They're asking themselves, *"Do these people understand what's at stake for us?"*

You communicate that through tone, rhythm, and empathy.

- Use plain, active language that sounds like people talking to people.
- Frame every example around the client's challenge, not your firm's performance.
- Add small signals of empathy with phrases such as *"You've told us…"* or *"We know how tight your timeframe is"* that show you're listening.
- End sections with forward-looking lines that turn your experience into their advantage: *"How we applied that lesson here…"*

Emotion connects the dots that logic can't. It's the reason a reviewer may remember your story long after all the scores are tallied.

Pulling the layers together

These three layers aren't steps you finish one at a time. They overlap, like transparent sheets that only form a full picture when stacked together.

- The machine layer keeps you *eligible.*
- The human layer keeps you *readable.*
- The emotional layer keeps you *memorable.*

Get those working in sync, and your proposal starts to feel inevitable. It becomes the kind of submittal that makes reviewers say, *"This one just makes sense."*

Producing the proposal first draft with AI

You've been waiting patiently for this part, so roll up your sleeves. We're about to use your AI Assistants to produce real, useable content.

However:

Before using any AI tool, confirm that your firm's data-privacy standards are met. Redact or generalize client identifiers and proprietary details unless your organization's enterprise account specifically allows their use in a secure, closed environment.

It's not the least bit necessary to surrender either creativity or control to get real help from AI. In fact, using these tools can enhance both. Used well, AI feels less like a machine and more like an eager junior teammate: they're enthusiastic as all get out, lightning fast on the keyboard, and good with patterns. But they know little or nothing about you, your business, and your clients, and they're easily distracted unless you keep them anchored.

The goal here isn't to automate writing; it's to *co-write* efficiently and confidently. That begins with a simple rhythm that guides every project:

Draft → Distill → Decide

Before we get into that loop, let's talk about how you *prepare the conversation* that makes AI useful in the first place.

Preparing the conversation

Before you ever ask AI to "write," you have to tell it who it is and where it's standing. A clear, grounded briefing is the difference between an output that sounds right and one that sounds either like everyone else's or incoherent gibberish.

AI writes best when it knows the terrain.

Think of it like onboarding a new proposal coordinator. You wouldn't toss them a blank RFP and say, "Go for it." You'd walk them through the context, your expectations for them, your tone, your past work, and what's at stake. AI needs that same orientation every time you start a new pursuit. Investing some time sketching these points will make all the difference in the world.

1. Set the role and relationship

Clarify what you expect AI to do, and what you don't. Define the job, the skill set, and how its "voice" will blend with yours.

Again, treat the model as an eager junior teammate: fast, pattern-savvy, but easily distracted.

2. Define the context

What kind of pursuit is this? Who's on the other side of the table? Each element of context anchors the AI to your reality. Sketch out where the proposal sits in the relationship. Is this a new client or an existing one? Is it a competitive shortlist or contract renewal?

Good outputs begin with good orientation.

Include audience cues. Are they engineers, city staff, or corporate executives? Name the stage you're at. Are you in early ideation, mid-draft, or final polish? The clearer you are, the more grounded its voice will be.

3. Shape tone and voice before you begin writing

Specify the kind of tone and posture you want. Formal? Collaborative? Friendly? Then supply background materials that define that tone and your personality. Upload previous write-ups, positioning notes, project examples, or research summaries and let the AI know what it's looking at and what you want it to pay attention to.

The model can mimic tone, but only if you give it something genuine to mimic.

Here's an example of what that conversation might look like:

Prompt example:

"You are an expert in conceptualizing and drafting proposals for professional architectural services, with particular expertise in municipal projects. Our small (35-person) architecture firm would like your assistance as we update a proposal for a community recreation center for <City, State>. The audience is a city selection panel with a mix of city counselors, technical staff, and citizen reviewers. We've already developed a sustainability message and gathered examples from three relevant projects (attached here). Please draft a 150-word project approach statement written in a confident, plain-spoken voice. Avoid jargon. Focus on collaboration with the city and our experience balancing recreation needs with long-term maintenance costs."

That level of context up front will save you many *hours* and much frustration later. AI writes best when it knows the terrain.

Reminding the model where it's been

Even after you've written the perfect setup, AI can lose track as the conversation grows. That's normal. It doesn't "remember" like a person; it only knows what's visible in your current chat or document window.

Remember, You're best to treat it like a brilliant but slightly forgetful colleague.

1. Why AI needs a recap

Think of each session as starting fresh. If the conversation runs long, parts of your earlier context fade from its active memory. When the thread drifts, quality drifts with it. A quick recap can avoid hours of post-editing.

2. How to reinforce context mid-stream

Drop short reminder prompts throughout the exchange. These gently re-anchor tone, context, and priorities without halting momentum. Examples:

- *Please remember that we want to keep the same tone we used in the executive summary; conversational but professional.*
- *You already know our main project message: reliability through partnership. Please review the positioning document we shared earlier and keep that front and center.*
- *Refer back to the sustainability examples we used earlier in this session.*

Reinforcing the context in which you're working midstream keeps the quality from drifting and helps maintain a steady voice across all sections. You don't need to re-train the model every time. Just remind it who you are, what you're doing, what you've decided, and what you're asking it to do.

And, while the "please and thank you" aren't required, I find that using a courteous tone keeps me in the mindset of working with a colleague, rather than a machine. If you think of your AI partner as a capable multitasker with an occasionally spotty, short-term memory, you'll handle it right.

3. Keep a short-term memory log

To prevent confusion across versions, copy each refined section out of the chat window and into your central draft. Label each iteration clearly (v1, v2, v3) in your shared pursuit folder.

Reinforcing context mid-stream keeps quality from drifting and maintains a steady voice across sections. Over time, this rhythm of reminding becomes second nature. It's a normal part of the collaboration between human and machine.

Okay, let's get this proposal underway!

Step 1: Draft. Begin with a strong setup

Once the context is in place, you're ready for a first draft. The goal here is speed, not polish. Let AI create the clay you'll shape later. A good drafting prompt usually answers three things *what, for whom, and how.*

Component	Example
What	*Write a 200-word project description introducing our approach to storm-water management.*
For whom	*This is for a county facilities group with limited technical background.*
How	*Keep it confident and readable, focusing on results rather than process.*

A weak prompt would be short, with little context and direction: *"Write a proposal section about flood control."*

You'll make the prompt stronger, the more background and expectation you provide: *"Write a 200-word introduction for a proposal about flood control in the County of <____>. The audience is that County's municipal review board, the voting patterns of which you have already researched. Use plain language to explain our experience designing systems that manage runoff and protect neighborhoods. Use the examples we've provided to you to create that explanation. Include a short discussion of the project we completed in <City> to highlight our approach."*

The better you brief, the better the draft. You can't expect magic from a half-sentence because this process works like any other collaboration: the care you put in upfront sets the quality for everything downstream.

Step 2: Distill. Turn the drafts into direction

Now you shift from *creator* to *editor*. Read what the model gave you and ask:

- *Does this capture what we want to say?*
- *Does it sound like us?*
- *Is it too wordy, too generic, or missing a client-specific thread?*

Then set about to shape it. Tighten, rearrange, and simplify. It's also a good idea to loop small sections back to AI for refinement. Challenge it when the output has not been what you expected or need.

- *Summarize this paragraph into two shorter sentences.*
- *Keep the same content but replace the buzzwords with plain speech.*
- *Make this sound warmer and more direct.*
- *We realize that this project (description attached) would add more credibility to the section on sustainability. Please include reference to it in that section.*

Expect to go through several iterations, with each round distilling a little more clarity. In some instances, I've found that the first draft is perfect. In others, it has required three or four passes to get it right. The key is to keep it moving forward and avoid endless editing. Think of it as progressive sculpting: the form emerges as you go.

Data & document tip

Trying to keep track of the evolving conversation thread in the AI window can be difficult. They become very long, very quickly and, with supplemental questions, prompts, and clarifications, you can easily lose the continuity of your draft.

Each time the AI bot returns a new section of the document you're working on, copy and paste it into whichever word processing software you're using. Then, keep one central working draft in your firm's shared drive or pursuit folder. Label each AI iteration clearly (v1, v2, etc.) so your team can track changes and avoid losing the definitive version.

Step 3: Decide. Blend the two intelligences into the final document

Here's where your team earns its keep. In *Chapter 8*, we introduced the idea of "two intelligences"—machine speed and human sense. This is the point where they intersect. AI has already provided you with a first draft, and you've reviewed and edited it. In this phase, AI has some additional strengths that you should exploit:

- Checking consistency and grammar
- Catching and re-wording repetitive phrases
- Testing alternate openings or short summaries

But the final decisions must always be left to the people:

- *What do we emphasize for this specific client?*
- *Does the tone match our relationship with them?*
- *Does it support the pursuit strategy we agreed on earlier?*

A proposal succeeds when it balances efficiency with empathy. The AI helps build the frame, but your judgment fills it with character and truth. When a section reads smoothly *and* feels genuine, stop. Over-editing with AI can polish away your personality just as surely as under-editing leaves it rough. The sweet spot is clear, concise, and human-sounding.

The loop in practice

Draft → *Distill* → *Decide* isn't just a process. It's a rhythm that keeps tempo with your efficient collaboration. With time, it becomes second nature: you brief the AI assistant carefully, let it explore, then the people return to fine-tune and approve. Every round strengthens the partnership.

You'll quickly begin to sense where AI accelerates your work and where your own experience must steer. That is the modern proposal engine, fast, grounded, and human-led.

As one senior marketer said after trying this process, *"It's like adding another team member who types faster and has instant access to more information than anyone I know. I just have to keep it reminded of what really matters."*

Now that we've built a shared writing rhythm, the next step is to **see it visually.** In the next section, we're going to explore visual frameworks and layout logic. In other words, how to use design tools and simple structure checks to make your proposals as easy to *look at* as they are to *read.*

***

Visual frameworks and layout logic

A great proposal doesn't just read well, it *looks* like it's been thought through. Your reviewers decide a lot in the first few seconds their eyes land on a page and those vital first impressions count for a great deal. If the structure looks crowded or disjointed, the content already has an uphill climb.

The purpose of visual frameworks (you can also call it graphic design) is not decoration, it's *navigation.* A well-designed layout gives every reader, human *and* digital, the same quick orientation: *"Here's what I'm looking at, here's why it matters, and here's where to go next."*

While this isn't intended to be a lesson in graphic design, it's important that you see real examples of good and bad layouts. Each of these illustrations are from real proposals from real firms, although they've been modified to mask any identifying features. Further, all these example pages are reproduced in grayscale to emphasize the visual design structure instead of firm-specific branding colors.

Think like an architect, design like an editor

The process of designing your proposal layout mirrors architecture:

- Foundation = clear hierarchy (consistent headings and structure)
- Framing = logical flow and alignment
- Finish = tone, color, and restraint

Your layout should quietly say, *"we think in systems."*

Example: How the reader's eye scans a page

A reader never starts at the beginning and then reads straight down a page. They scan across it in a fast Z-pattern, landing first on major headings, then jumping diagonally to the next focal area before finishing on key messages at the lower right. A clear visual hierarchy guides that motion, allowing readers to find structure and meaning in seconds rather than minutes.

This diagram shows what reviewers' eyes tend to do when they open a page, either printed or on-screen.

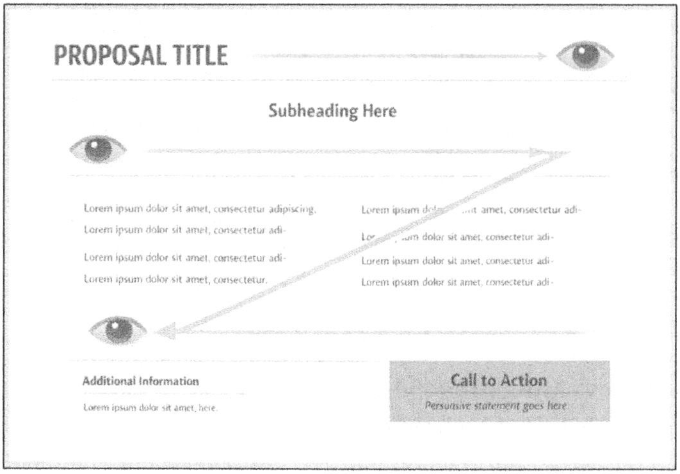

Figure 9.1 a *Effective layouts anticipate this pattern by placing primary messages, including titles, subheads, and calls-to-action, along the natural Z-path. Supporting text and data fit in the calm zones between those anchor points. This alignment makes proposals feel lighter, more modern, and easier to trust.*

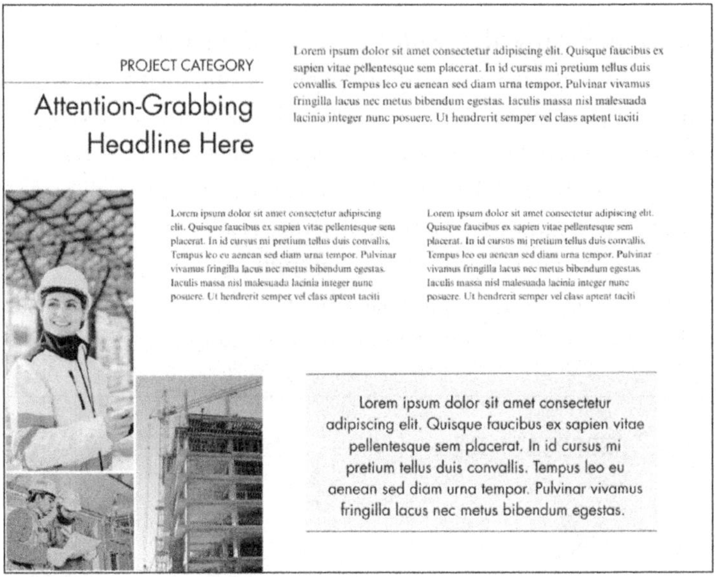

Figure 9.1 b *Example of effective visual eyeflow in a real proposal layout. Note how the headline, imagery, and text hierarchy guide the eye naturally without the need for directional signals.*

Legacy vs Modern proposal layout

The habits of both page designers and readers have evolved and the "art of information transfer" is decidedly different in the digital era. Here is how the characteristics of modern versus legacy layouts compare, Followed by examples of both.

Element	Legacy Layout (Print-Style)	Modern Digital Layout
Density	Full text columns, minimal white space	1½-column flow with breathing room

Element	Legacy Layout (Print-Style)	Modern Digital Layout
Hierarchy	Indistinct: headings same size as body	Clear levels (H1, H2, H3) plus bold sub-lines
Emphasis	Underlines/italics/borders	Short callouts, shaded quotes
Navigation	Page numbers only	Section tabs / linked Table of Contents
Visuals	Infographic art added last	Visual rhythm planned from the outset
Accessibility	PDF-only	Readable on mobile, tagged headings

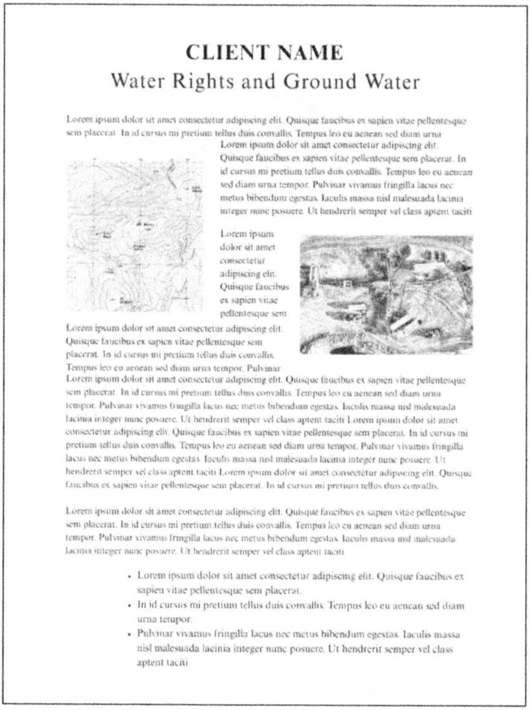

CLIENT NAME

Water Rights and Ground Water

Lorem ipsum dolor sit amet consectetur adipiscing elit. Quisque faucibus ex sapien vitae pellentesque sem placerat. In id cursus mi pretium tellus duis convallis. Tempus leo eu aenean sed diam urna

Lorem ipsum dolor sit amet consectetur adipiscing elit. Quisque faucibus ex sapien vitae pellentesque sem placerat. In id cursus mi pretium tellus duis convallis. Tempus leo eu aenean sed diam urna tempor. Pulvinar vivamus fringilla lacus nec metus bibendum egestas. Iaculis massa nisl malesuada lacinia integer nunc posuere. Ut hendrerit semper vel class aptent taciti

Lorem ipsum dolor sit amet consectetur adipiscing elit. Quisque faucibus ex sapien vitae pellentesque sem

Lorem ipsum dolor sit amet consectetur adipiscing elit. Quisque faucibus ex sapien vitae pellentesque sem placerat. In id cursus mi pretium tellus duis convallis. Tempus leo eu aenean sed diam urna tempor. Pulvinar

Lorem ipsum dolor sit amet consectetur adipiscing elit. Quisque faucibus ex sapien vitae pellentesque sem placerat. In id cursus mi pretium tellus duis convallis. Tempus leo eu aenean sed diam urna tempor. Pulvinar vivamus fringilla lacus nec metus bibendum egestas. Iaculis massa nisl malesuada lacinia integer nunc posuere. Ut hendrerit semper vel class aptent taciti Lorem ipsum dolor sit amet consectetur adipiscing elit. Quisque faucibus ex sapien vitae pellentesque sem placerat. In id cursus mi pretium tellus duis convallis. Tempus leo eu aenean sed diam urna tempor. Pulvinar vivamus fringilla lacus nec metus bibendum egestas. Iaculis massa nisl malesuada lacinia integer nunc posuere. Ut hendrerit semper vel class aptent taciti Lorem ipsum dolor sit amet consectetur adipiscing elit. Quisque faucibus ex sapien vitae pellentesque sem placerat. In id cursus mi pretium tellus duis convallis.

Lorem ipsum dolor sit amet consectetur adipiscing elit. Quisque faucibus ex sapien vitae pellentesque sem placerat. In id cursus mi pretium tellus duis convallis. Tempus leo eu aenean sed diam urna tempor. Pulvinar vivamus fringilla lacus nec metus bibendum egestas. Iaculis massa nisl malesuada lacinia integer nunc posuere. Ut hendrerit semper vel class aptent taciti

- Lorem ipsum dolor sit amet consectetur adipiscing elit. Quisque faucibus ex sapien vitae pellentesque sem placerat.
- In id cursus mi pretium tellus duis convallis. Tempus leo eu aenean sed diam urna tempor.
- Pulvinar vivamus fringilla lacus nec metus bibendum egestas. Iaculis massa nisl malesuada lacinia integer nunc posuere. Ut hendrerit semper vel class aptent taciti

Figure 9.2 a: (previous page) *Example of a legacy layout. Note how the dense blocks of text and the absence of a visual hierarchy force the reader to read every word as they dig for structure and key information. This dramatically slows overall comprehension and often leads to the dreaded TL;DR (Too Long; Didn't Read)*

Figure 9.2 b: *Example of a modern layout. The clean grid, defined hierarchy, generous white space, and a single conceptual image demonstrate how modern proposals guide reading with economy and emotional resonance.*

Building a visual rhythm

You've already established a writing rhythm with the *Draft* → *Distill* → *Decide* technique. Now we're going to add a *visual rhythm* to create a balance between consistency and visual cueing.

1. **Consistency**
    - Headings and subheads follow a recognizable and reliable hierarchy (H1–H3).

- Caption and table formats repeat exactly across page spreads.

2. **Cueing**
   - Use color, icons, and spacing as directional signage.
   - Give visual priority to content types. For example:
     - Problems = warm tone
     - Solutions = neutral tone
     - Results = cool tone (calming confidence)

3. **Breathing space**
   - Surround heavy text blocks with generous white space; it shows confidence and makes digital reading effortless.

Example: Grid and white-space balance

As generative AI streamlines proposal assembly, designers and marketers are freed to move beyond static templates. The resulting flexibility invites you to create layouts that are structured but adaptive, built with a professional graphic designer, from modular components, or even AI-generated image pairings.

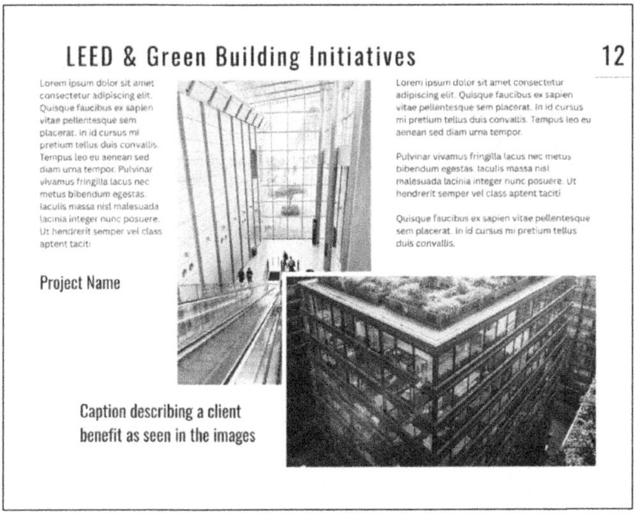

Figure 9.3 (previous page): Modern Layout: *This example demonstrates how a flexible block composition can maintain clarity and rhythm without relying on a rigid grid. Generous white space defines the page's hierarchy, creating natural zones for text, imagery, and project identification. The varied scale of the image blocks adds energy and movement, while consistent typographic alignment ensures order. Layouts such as this reflect how designers can now prototype and test multiple arrangements quickly with AI-assisted tools, using algorithms to balance content density, hierarchy, and impact.*

The 10-second test

You can teach your team this quick visual check before final submission. Within the first ten seconds of your eyes landing on the screen or page, what stands out first and what is the sequence of seeing the rest of the content? Show the layout to someone who hasn't had any involvement with the proposal and ask what they see first.

- Headline?
- Introductory sentence?
- Callout quote?
- Body paragraph text?
- Next subheading?

If the quick glance lands on the substantive ideas first, your design is working. If you find your eyes wandering over the page, not sure what to settle on, or only noticing shapes, line, or colors, refine the design again.

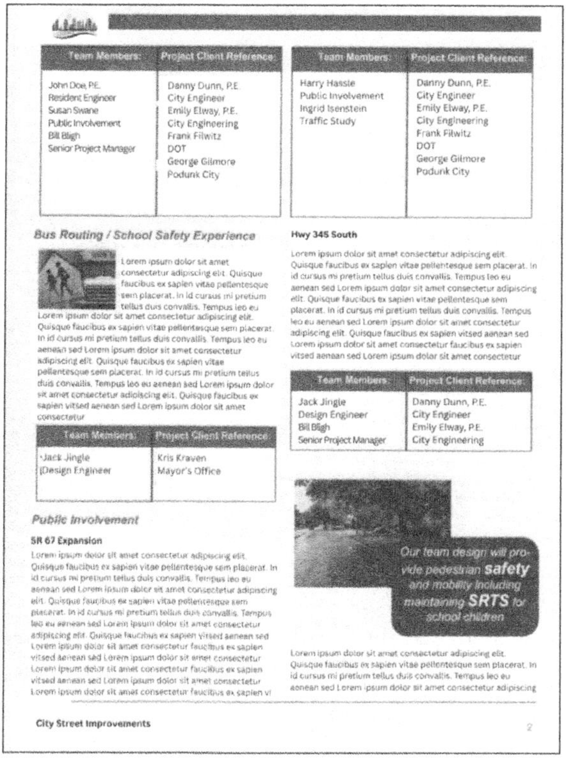

Team Members:	Project Client Reference:
John Doe, P.E. Resident Engineer Susan Swane Public Involvement Bill Bligh Senior Project Manager	Danny Dunn, P.E. City Engineer Emily Elway, P.E. City Engineering Frank Filwitz DOT George Gilmore Podunk City

Team Members:	Project Client Reference:
Harry Hassle Public Involvement Ingrid Isenstein Traffic Study	Danny Dunn, P.E. City Engineer Emily Elway, P.E. City Engineering Frank Filwitz DOT George Gilmore Podunk City

Bus Routing / School Safety Experience

Lorem ipsum dolor sit amet consectetur adipiscing elit. Quisque faucibus ex sapien vitae pellentesque sem placerat. In id cursus mi pretium tellus duis convallis. Tempus leo eu Lorem ipsum dolor sit amet consectetur adipiscing elit. Quisque faucibus ex sapien vitae pellentesque sem placerat. In id cursus mi pretium tellus duis convallis. Tempus leo eu aenean sed Lorem ipsum dolor sit amet consectetur adipiscing elit. Quisque faucibus ex sapien vitae pellentesque sem placerat. In id cursus mi pretium tellus duis convallis. Tempus leo eu aenean sed Lorem ipsum dolor sit amet consectetur adipiscing elit. Quisque faucibus ex sapien vitsed aenean sed Lorem ipsum dolor sit amet consectetur

Team Members:	Project Client Reference:
Jack Jingle Design Engineer	Kris Kraven Mayor's Office

Public Involvement

SR 67 Expansion

Lorem ipsum dolor sit amet consectetur adipiscing elit. Quisque faucibus ex sapien vitae pellentesque sem placerat. In id cursus mi pretium tellus duis convallis. Tempus leo eu aenean sed Lorem ipsum dolor sit amet consectetur adipiscing elit. Quisque faucibus ex sapien vitae pellentesque sem placerat. In id cursus mi pretium tellus duis convallis. Tempus leo eu aenean sed Lorem ipsum dolor sit amet consectetur adipiscing elit. Quisque faucibus ex sapien vitsed aenean sed Lorem ipsum dolor sit amet consectetur faucibus ex sapien vitsed aenean sed Lorem ipsum dolor sit amet consectetur Lorem ipsum dolor sit amet consectetur faucibus ex sapien vitsed aenean sed Lorem ipsum dolor sit amet consectetur Lorem ipsum dolor sit amet consectetur faucibus ex sapien vi

Hwy 345 South

Lorem ipsum dolor sit amet consectetur adipiscing elit. Quisque faucibus ex sapien vitae pellentesque sem placerat. In id cursus mi pretium tellus duis convallis. Tempus leo eu aenean sed Lorem ipsum dolor sit amet consectetur adipiscing elit. Quisque faucibus ex sapien vitae pellentesque sem placerat. In id cursus mi pretium tellus duis convallis. Tempus leo eu aenean sed Lorem ipsum dolor sit amet consectetur adipiscing elit. Quisque faucibus ex sapien vitsed aenean sed Lorem ipsum dolor sit amet consectetur faucibus ex sapien vitsed aenean sed Lorem ipsum dolor sit amet consectetur

Team Members:	Project Client Reference:
Jack Jingle Design Engineer Bill Bligh Senior Project Manager	Danny Dunn, P.E. City Engineer Emily Elway, P.E. City Engineering

Our team design will provide pedestrian ***safety*** *and mobility including maintaining* ***SRTS*** *for school children*

Lorem ipsum dolor sit amet consectetur adipiscing elit. Quisque faucibus ex sapien vitae pellentesque sem placerat. In id cursus mi pretium tellus duis convallis. Tempus leo eu aenean sed Lorem ipsum dolor sit amet consectetur adipiscing

City Street Improvements 2

Figure 9.4 a: Legacy Layout. *Dense text, repetitive tables, uneven visual hierarchy, and no white space make this page difficult to interpret at a glance. A reader must hunt for relationships between projects, teams, and outcomes, failing the "10-second test" of first-impression clarity. The design reflects a manual, assembly-line approach where content was placed section by section rather than orchestrated for visual communication.*

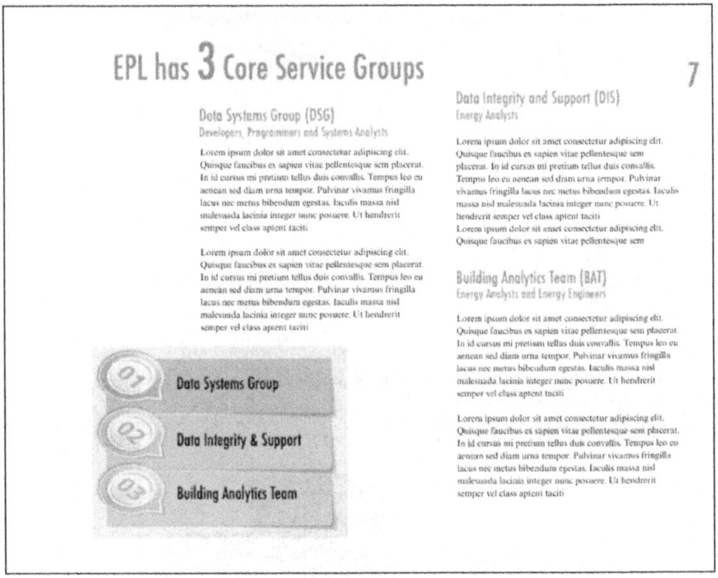

Figure 9.4 b: Modern Layout. *This layout passes the "10-second test" by surfacing hierarchy and meaning immediately. Clear typography, modular grouping, and simple iconography communicate structure before a reader has even begun line-by-line reading. Such clarity is now accelerated by AI-assisted formatting and content-generation tools that can identify key message groupings, prioritize information, and suggest high-impact visual organization. This allows marketers to focus on narrative intent rather than mechanical layout.*

Encourage the reviewers and quality control members of your team to use this test on every page, especially digitally formatted ones.

Applying AI in the visual stage

By uploading your layout design, AI-assisted design tools can now check legibility and accessibility automatically.

- Layout evaluation: Platforms that score contrast, spacing, and read-time.

- Auto-tagging: Assistive tags for screen readers (critical for public-sector PDFs).
- Visual balance suggestions: Analyze color usage and density.
- Preview modes: Simulate mobile, tablet, and desktop to catch flow issues early.

But the same rules always apply. AI can propose a design, but the human designer must always make the final decision. People still have to judge what *feels right.*

Having explored how readability, hierarchy, and visual pacing affect a viewer's all-important first impression, it's useful to pause and translate those lessons into practice. This simple check distills the design and communication principles we've noted in the examples into an evaluative tool for marketers.

Each item can serve as the basis of a prompt when reviewing proposal drafts, whether it's a manual prompt to remind your team, or a written prompt to initiate the assistance of AI layout engines. Either way, this quick review will help ensure your final proposal document communicates its story clearly, credibly, and within that critical ten-second window.

Quick review for AI-assisted layout

1. Upload your draft proposal text and existing brand template.
2. Ask AI to suggest spacing, font scaling, and image placement.
3. Check automated readability or accessibility feedback, then adjust accordingly. (Many AI-assisted layout tools now highlight low-contrast text, missing headings, or other elements that can impede readability. I've provided a list at the end of this chapter.)
4. Have the human designer review the document for tone and brand accuracy.
5. Export the final draft in both print and screen layouts.

Each of these steps keeps AI focused on mechanical refinements, which is its strength, while letting you focus on storytelling and consistency.

Bringing writing flow and visual flow together

When your narrative rhythms and layout rhythms align, reviewers feel it immediately. Sentences guide the eyes, while visual design guides the attention. That's how proposals start to "read themselves," even in motion on a tablet.

The goal is holistic readability, which is easy on the software, easy on the people, and emotionally resonant in split-second impressions. With the writing and visual systems in sync, it's time for the final human layer: the three conscious read-throughs that turn good drafts into trusted communication with *Tone, Logic, and Empathy*.

In the next section, we'll learn how teams reclaim and polish the document. Even the smartest AI and the cleanest layout still rely on human warmth to make the proposal *land.*

Human edit checkpoints

Even the most advanced AI can only approximate good judgment. The real polish, the part reviewers can *feel*, comes from the refinement that people give to your document. Before sending any proposal to print or upload, you need to plan three deliberate human read-throughs. Each one has a distinct purpose and keeps the final product grounded in integrity and brand truth.

1. **Tone. Does it sound like us?**

   Your firm's voice is one of its most valuable assets.

   - Does the proposal speak with your usual balance of confidence and humility?
   - Are there phrases that sound automated, inflated, or "off tone"?
   - Are empathy cues intact ("you," "your project," "our partnership")?

Read the key pages aloud, especially introductions, approach summaries, and cover letters. If a sentence does not sound natural when spoken, polish it until it does.

2. **Logic. Does it flow in the manner in which people think?**

   Check for structural clarity.

   - Does each section lead naturally into the next?
   - Are transitions between team descriptions, project examples, and outcomes smooth?
   - Do key data points (dates, roles, metrics) appear in consistent formats?

A second reader, ideally someone not directly involved in the pursuit, should be able to skim and retell your narrative in order. If they can't, refine the logic.

3. **Empathy. Does every line show that you understand what the client is needing?**

   The final pass is not about grammar. It's about intent.

   - Do you restate the client's goals using their own language? (but without simply cutting and pasting verbatim)
   - Have you emphasized outcomes over effort?
   - Does the layout give enough emphasis and visual "breathing room" to the ideas the client cares about most?

Empathy edits are where proposals move from technically sound to emotionally persuasive.

Putting it in rhythm

Treat these three passes as a finishing rhythm: *Tone* → *Logic* → *Empathy*. Each one sharpens a different dimension of trust. Together, they ensure that what the AI helped build now speaks unmistakably in your firm's own voice. When the document reads clearly, moves logically, and sounds like people who care, that's when the proposal stops being a submittal and starts being a conversation.

Example: Before and after

The difference between *traditional* and *AI-augmented* proposal pages is rarely about new ideas. It's about how quickly and coherently those ideas show up visually. In this brief example, both versions make use of the same content: a one-page project summary describing a municipal recreation center. Only the design and layout workflow change.

Evergreen Community Recreation Center

Reconnecting a neighborhood through sustainable design

When the City of Evergreen set out to renew its aging 1970s recreation facility, the goal extended beyond new walls and roofs. The city wanted a building that could adapt, conserve resources, and invite residents back in—from early-morning lap swimmers to after-school basketball leagues. Our team's challenge was to reshape a beloved but limited structure into a modern, low-carbon community hub that felt open, efficient, and innately local.

Understanding the Neighborhood

Early engagement confirmed the community's desire for something welcoming yet responsible. The existing 48,000SF structure was energy-intensive and visually disconnected from its park setting. Our architects hosted a two-day design charrette with nearly sixty participants, from city staff to neighborhood volunteers. Their input guided every major decision—from orientation of the new glazing walls to the multipurpose studio that doubles as a polling site during elections.

Design Response

The renovated complex introduces a daylight-harvesting roof form and a flexible interior spine linking the gymnasium, natatorium, and classroom zone. Structural retrofits preserved 70 percent of the existing steel frame, reducing embodied carbon by an estimated 38 percent. Local materials—Douglas-fir glulam beams and aggregate from a nearby quarry—lowered transport impact and echoed the Northwest's natural palette.

Figure 9.5 a: *Traditional (manual) layout. Text was pulled from a legacy template and formatted manually. Headings are uneven, captions vary, and layout spacing was adjusted by eye. Reviewers must read line by line to locate the main points.*

Common traits:

- Dense paragraph block after block
- Irregular spacing between titles and images
- Inconsistent typography and tone
- No instant sense of hierarchy. The eye much search rather than follow

At a glance, the page looks *complete* but feels heavy. Readers experience fatigue before comprehension.

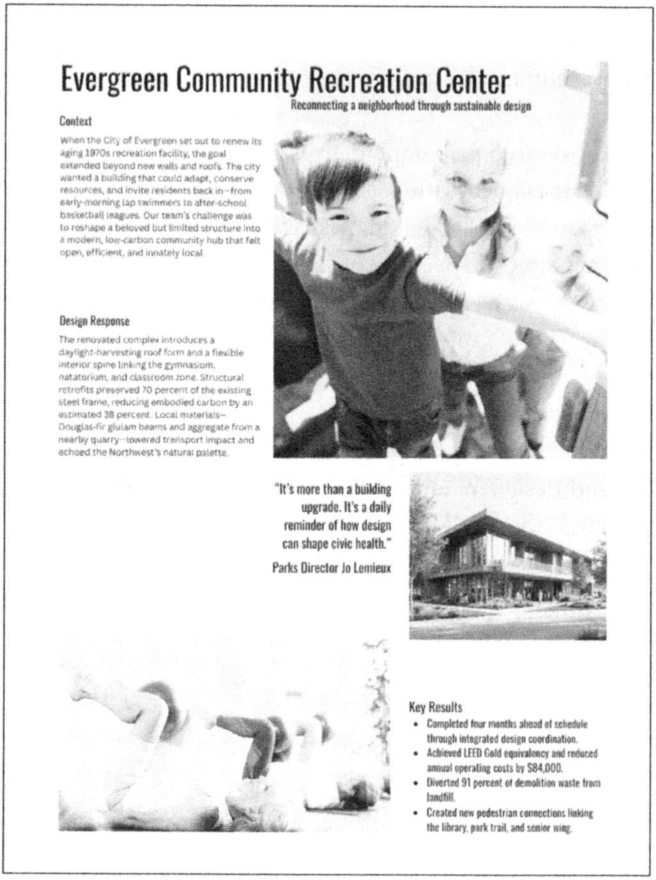

Evergreen Community Recreation Center

Reconnecting a neighborhood through sustainable design

Context

When the City of Evergreen set out to renew its aging 1970s recreation facility, the goal extended beyond new walls and roofs. The city wanted a building that could adapt, conserve resources, and invite residents back in—from early-morning lap swimmers to after-school basketball leagues. Our team's challenge was to reshape a beloved but limited structure into a modern, low-carbon community hub that felt open, efficient, and innately local.

Design Response

The renovated complex introduces a daylight-harvesting roof form and a flexible interior spine linking the gymnasium, natatorium, and classroom zone. Structural retrofits preserved 70 percent of the existing steel frame, reducing embodied carbon by an estimated 38 percent. Local materials—Douglas-fir glulam beams and aggregate from a nearby quarry—lowered transport impact and echoed the Northwest's natural palette.

"It's more than a building upgrade. It's a daily reminder of how design can shape civic health."

Parks Director Jo Lemieux

Key Results

- Completed four months ahead of schedule through integrated design coordination.
- Achieved LEED Gold equivalency and reduced annual operating costs by $84,000.
- Diverted 91 percent of demolition waste from landfill.
- Created new pedestrian connections linking the library, park trail, and senior wing.

Figure 9.5 b (previous page): *AI-Augmented Layout. The same, albeit reduced text, (less is more) and image were imported into an AI-assisted layout tool. Additional, life-style images were added for emotional appeal. Spacing, font scaling, and call-out order were suggested algorithmically, then refined by a human editor. Within seconds, the tool balanced white space, recast subheads into consistent hierarchy, and generated alternate compositions optimized for screen readability. The human designer selected and polished the best variant.*

Resulting traits:

- Consistent header progression (H1–H3) guiding the scan path
- Auto-balanced paragraph density and margin spacing
- Call-outs aligned with key client outcomes
- Clear anchor image linking visually to section title

The net effect is speed + cohesion + ease of reading. Reviewers intuitively know where to look first, and the firm's message lands faster.

The core lesson

AI didn't write anything new. It simply accelerated what good editors and designers already do, which is to provide structure, balance, and priority. The technology made the mechanical decisions so people could make the *story* decisions.

In short, machine speed sets the frame, while human sense builds the trust. When both work in rhythm, proposals read as if they were effortless.

Recapping the modern proposal formula

By now, you've seen that the modern proposal isn't just a document. It's a living, readable system, constructed through rhythm, clarity, and empathy. Each layer, from the technical to the visual to the emotional, shapes how readers experience credibility and care.

- *Machine-Readable* made sure your work could be *found.*
- *Human-Readable* ensured it could be *followed.*
- *Emotionally Persuasive* made it *felt.*

Together, those layers create proposals that don't just inform, they *flow.* And when AI enters the process, not as an author, but as a rhythmic collaborator, you gain time and cohesion without losing authenticity. The *Draft* → *Distill* → *Decide* loop becomes your operating tempo. It's fast enough to keep momentum, but deliberate enough to keep its humanity.

The best firms now treat every pursuit as an experiment in clarity by blending algorithmic precision with editorial warmth. That's not the end of transformation, it's the foundation for what comes next.

From proposals to presence

In your proposals, you control the client experience through page rhythm; how the eyes move and how stories resolve. But what happens once your story leaves the page? Screens, slides, and meetings now extend your narrative into real time. Your team's energy, posture, and pacing become the next "layout" clients read.

Chapter 10 will translate these same principles of rhythm, hierarchy, and empathy into *live communication.* You'll learn how to design presentations that move with the same cohesion your proposals now express. Each slide, voice cue, and transition contributes to one unified, confident experience.

AI Layout Tools in Practice

AI-supported layout design is no longer limited to specialized graphic-design environments. Many mainstream proposal and marketing platforms now embed intelligent layout features that can analyze composition, suggest hierarchy, and flag accessibility concerns.

Tools such as *Adobe Express, Canva, Microsoft Designer*, and *Figma with Autolayout and AI Assist plugins* provide automated alignment, spacing, and visual-balance options. *Acrobat Pro's accessibility checker* evaluates contrast and tag structure, while presentation-focused platforms like *Tome* and *Pitch* use AI to generate slide or page layouts from text prompts.

These tools don't replace design expertise, they accelerate it. For AE marketers, they serve as a bridge between technical content and clear visual communication, allowing design iteration and accessibility checks to happen early in the proposal process, where they can have the greatest strategic impact.

CHAPTER 10

Presentations 2.0: Energy across every screen

The new presentation reality

A proposal puts your ideas on paper; a presentation brings them to life. Today, that "in-person moment" rarely happens in a single room. Half the panel might be in city hall, another member on a laptop miles away, and one more replaying your recorded answers on their own schedule. And they're all judging in different ways.

Welcome to the hybrid room. It's a shifting mixture of in-person, virtual, and asynchronous encounters where the definition of "audience" no longer fits a single moment.

Hybrid panels, digital interviews, asynchronous scoring

The new presentation reality is built around fractured attention. Reviewers toggle between screens, spreadsheets, and chat windows. Some hear you live, others judge you later through a replay. Camera framing, lighting, and bandwidth suddenly rank alongside design credentials in shaping first impressions.

The real challenge, though, isn't technical competence, it's *emotional coherence.* Can your team feel like one confident, grounded group, even if they're not physically together and the audience is scattered?

Firms that once rehearsed around a conference table now rehearse across screens. Micro-delays and camera angles reveal things we never used to see: darting eyes, mismatched energy, the quiet colleague who fades into the grid. Each square tells a part of the whole story before a single word has been spoken. And for selection panels dealing with digital fatigue, those quick signals often make the difference between who feels trustworthy and who's on the reject list.

Composure = Credibility

Hybrid environments reward presence, not perfection. When technology mediates eye contact, the smallest cues, such as your posture, tone, and the momentary pause before answering all combine to broadcast either composure or the lack of it. And composure reads as competence. Audiences may forgive a fuzzy connection or a stumble in phrasing, but they notice the person who stays calm, centered, and generous through it all.

That's where preparation has changed. Old-style rehearsals focused on *what* you said. Modern preparation adds a new layer, *how your presence translates through glass.* The goal isn't performance theater; it's digital empathy, in which you project steadiness, curiosity, and warmth at pixel-range distance. The room may be virtual, but credibility is physical. It's carried in your breathing, pacing, and tone.

In the coming sections, we'll unpack how to balance energy and information, how AI rehearsal tools can give you feedback without flattening authenticity, and how to design both slides and stories that move with intention.

For now, remember that the modern presentation is not a meeting. It's a multilayered moment of impression management. And mastering that begins with the presence and composure of your team that transmits through every screen.

Energy vs Information

In every presentation, two kinds of energy move through the room. One comes from the content. The other comes from the people who deliver it. When both are aligned, the message feels alive. When they aren't, the audience may hear your words but still lose interest.

Most presentations don't run out of information. They run out of energy.

Data energy and human energy

Data energy comes from the thoroughness of your preparation. It's the confidence that your material is solid, organized, relevant, and filled with the answers to their questions. This type of energy translates into steady pacing, simple graphics, and concise explanations.

Human energy comes from tone and connection. It's what the audience reads between the lines. In other words, it's how much you care, how present you are, and how aware you seem of their reactions. Human energy makes information land as meaning, not just data.

Too many presenters try to rely almost entirely on the energy that the data and information provide. But it's the balance between these two energies that defines every successful presentation. Too much data energy and you sound mechanical. Too much human energy and you risk sounding ungrounded. The goal is to find the point where clarity meets warmth.

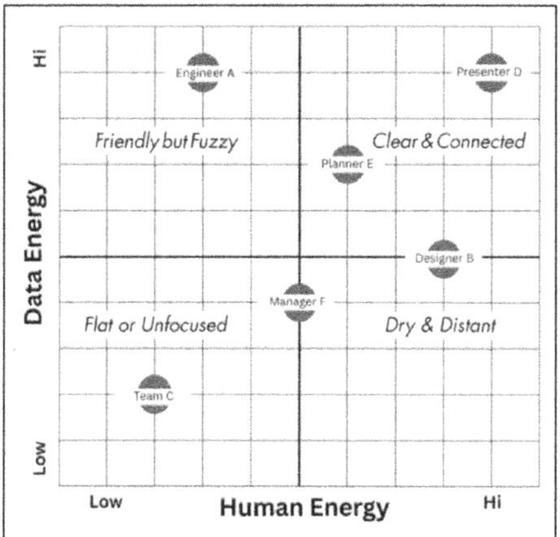

Balancing Data and Human Energy

Figure 10.1 (previous page): *Balancing data and human energy Presentations succeed when content confidence and human connection intersect. Too much of one without the other produces imbalance. The upper-right zone marks where clarity and warmth coexist.*

Why balance matters

Selection panels don't consciously map this out, but they respond to it instinctively. When a presenter is calm, concise, and responsive, the panel feels both informed and respected. That combination builds trust far faster than a rehearsed speech or a flood of slides.

When energy drops, attention does too. When energy spikes, credibility can wobble. The best presenters manage tempo the way designers manage white space: alternating focus and breathing room so that key messages stand out.

In hybrid or digital settings, this balance becomes even more visible. Viewers can sense exactly when enthusiasm turns forced or when a slide deck feels like a wall of noise. That's why feedback and rehearsal matter. They're not about memorizing lines but about finding the tempo that feels authentic on screen.

How to reset the balance

1. **Check for weight, not volume**

   If your slides carry the entire story, add more human context. If everything depends on charisma, bring the content back into focus.

2. **Match tone to tempo.**

   Speak slightly slower than you think you should. Let pauses carry confidence. A steady rhythm signals control more than speed does.

3. **Design slide rhythm to rest the mind.**

   Alternate detail slides with simple visual breaks. The change in texture revives attention the same way silence strengthens sound.

When data energy and human energy reinforce each other, your audience feels both informed and safe. They lean in because you project clarity without forcing it.

AI-and-human rehearsal loop

Before most presentations, teams rehearse to hear each other's timing and fix rough edges. Today, those sessions can be smarter. AI-based rehearsal tools now watch what an audience would see. They measure pace, phrasing, and even micro-expressions. Used well, they don't replace coaching, they extend it.

How these tools work

Rehearsal platforms analyze three layers of delivery:

- **Voice**: Identifies filler words, speech rate, pacing, and tone variation.
- **Face**: Reads engagement cues such as eye contact, expression variety, and focus.
- **Flow**: Tracks slide timing and audience attention patterns during shared presentations.

When used thoughtfully, these insights help a team balance tempo, maintain composure, and project confidence on screen.

Recommended AI rehearsal tools

The following list reflects a mix of free-tier and subscription tools. These are widely available commercial platforms as of 2026. Capabilities evolve quickly, so test them for fit and privacy before use.

Purpose	App Examples	Typical Output/ Insights
Speech analysis	Orai Yoodli PitchVantage VirtualSpeech	Counts filler words, analyzes pace and clarity, highlights strong or weak phrasing.

Purpose	App Examples	Typical Output/ Insights
On-camera feedback	Rehearsal Pro AI Swell Tavus AI Captions by Descript	Detects eye-contact consistency, posture, and tone; sometimes offers real-time scoring or sentiment report.
Slide + timing sync	PowerPoint Speaker Coach (Microsoft 365) Google Slides Rehearse with Me	Tracks speaking speed per slide, suggests pacing corrections, flags over-dense visuals.
Comprehensive rehearsal dashboards	Supernormal Synthesia Presenter View Veo AI Coach	Integrates audio, video, and attention metrics to simulate audience perception.

How to use them effectively

1. **Record your normal rehearsal.**
   Open your webcam and microphone, share your slides, and present the way you would live. Don't aim for perfection on the first run. Just capture your natural baseline.

2. **Upload the recording to your selected tool.**
   Or use its built-in recording feature. Most apps accept standard .mp4 videos and generate a dashboard or transcript automatically.

3. **Review key metrics.**
   Focus on:
   - Average speaking pace (words per minute)
   - Total "um/uh" count
   - Eye-contact percentage
   - Slide dwell time
   - Tone variance (energy curve)

4. **Refine one element at a time.**
   Re-rehearse concentrating on a single dimension. For example, slower pacing or stronger hand-offs between speakers.

5. **Re-record live if possible.**
   Some tools (*Yoodli, Orai*) can run concurrently during online meetings, offering real-time prompts. But use them sparingly since the purpose is feedback, not scripted delivery.

6. **Transition to a human review.**
   After the AI rehearsal, switch to a peer or mentor session to test warmth, clarity, and flow. While machines can flag habits, only people can judge authenticity.

Using data for presence, not performance

The numbers you get from the AI feedback should point you to awareness, not imitation. If the AI flags short eye contact, strengthen your connection to the content rather than forcing a stare into the lens. Authenticity grows when feedback clarifies *why* something feels off, not when it forces a robotic fix.

The best teams treat digital rehearsal as a *mirror*, not a mask. Data amplifies what's already there, including your tone, curiosity, and composure, so you can see it sooner and refine it faster.

Now that the team understands *how* to balance energy and measure delivery, the next step is to shape *what* they say, so your story sounds unified even across multiple voices.

Designing the verbal structure

Once your energy is balanced and your delivery feels anchored, the next step is to shape what you say. A clear verbal structure keeps your team aligned even when screens, time zones, and speaking styles differ. Think of the verbal structure as the frame that holds your collective story. It ensures the audience hears one argument, not eight personal versions of it.

From content outline to story frame

Most teams begin with a slide order. (Worse, still, many begin by simply copying the slides from a previous presentation.) But slides are merely containers. They aren't the narrative. What panels and selection reviewers actually hear is a sequence of logic and emotion. They are listening for why this project matters, why this team fits, and why the decision feels safe.

A reliable model for this is a mini-story arc that works across interviews and presentations alike: *Promise → Proof → Personality → Promise Returned*

1. **Promise**

Begin every segment, whether it's the project introduction or a specialized technical answer, by telling the panel what the value will be. In one crisp sentence, set the intent: *"Here's what we set out to solve,"* or *"Our design goal is to…"* This acts as the attention magnet. Listeners immediately know why they should care and how to follow your logic.

2. **Proof**

Next, show the concrete evidence. The proof stage grounds your promise with specifics that can consist of data, precedents, cost frames, images, or design logic. Keep this portion proportionate. Your audience engages best when you move from concept to evidence within a 30-to-60-second window. Too much proof without narrative momentum begins to sound like a data dump.

When slides carry detailed graphics, rehearse a single sentence that interprets each image: *"What you're seeing here is how the daylight study confirms the comfort goal we promised at the start."* A line like that bridges the promise and the proof.

3. **Personality**

Review panels remember people, not slide decks. Over my many years working with AE firms I've spoken with countless clients who can't remember the content of the presentation, but they clearly remember liking the team. Or the opposite.

To make that connection, inject a short human-interest story. This could be a previous challenge that you overcame, a personal anecdote about the community, or a moment of discovery, just to name a few examples.

This isn't a place to share a "fun fact." It's an opportunity to convey personal stake, care, and collaboration. Examples include:

- Adding a quick empathy note: *"During the site walk, a teacher mentioned how often the classrooms overheat by mid-day, so we're going to adjust the shading angle."*
- Sharing something personal: *"I grew up around the corner from this site, and I remember my friends and I riding our bikes past this building."*
- Pulling back the curtain on your internal process: *"When our structural engineer suggested an alternate column grid, it opened a daylight corridor we hadn't seen before. That moment reshaped how the whole team thought about circulation."*

Now the story has taken on a personal tone and the stakes you have in the project's success have gone up. No other competitor will be able to tell those stories. This will strengthen your connection with the client in ways that your expertise and experience never can.

4. Promise returned

Every section should close by circling back to the opening value. That's how segments link into a coherent pitch. Conclude briefly: *"So the result ties back to our original goal, which is comfort and orientation expressed through daylight."* This echo signals resolution and helps each speaker hand off seamlessly to the next.

Building continuity across speakers

In hybrid or multi-presenter sessions, clunky or uneven hand-offs are the biggest cause of rhythm loss. To prevent that:

- Draft short transition cues between individuals: *"I'm going to hand this off to Maya, who'll expand on how that comfort idea shaped the interiors."*
- Keep a verbal through-line anchored around three to five recurring project words or terms such as *adaptability, clarity,* or *community*. When every team member repeats one or two of these words naturally, the presentation sounds unified without sounding over-rehearsed.

For distributed teams, record one dry run and mark the spots where the natural overlaps occur. Then trim or adjust the order rather than forcing everyone into equal time blocks.

Reviewers will always notice and appreciate an organic and appropriate pacing more than any artificial equity of airtime.

Practicing the story

Before the final rehearsal:

1. Print or share your slide thumbnails, two per page.

2. Under each slide, write one line per speaker following the *Promise–Proof–Personality–Promise Returned* framework.

3. Read through the sequence out loud without showing the slides. If you lose the thread, it means your narrative spine isn't clear yet.

4. Each speaker keeps that one-line sheet handy during digital rehearsals. It functions as a light script outline without reducing natural tone.

Why this structure matters

Panels rarely score your delivery line by line. Instead, they look for and reward *coherence*. The *Promise-Proof-Personality-Promise Returned* pattern ensures that your presentation is coherent through rhythm and repetition, making it easier for listeners to recall your points later.

It also surfaces weak spots early. If your "*proof*" sections depend on jargon, you'll feel it when the pattern breaks. If your "*personality*" moments feel thin, you'll see which voices need a more human contribution.

When teams internalize this framework, the result isn't a memorized script. It's a conversation held within clear boundaries. And that clarity is apparent in every slide and across every screen.

In the next section, we're going to translate this verbal rhythm into the pacing, hierarchy, and elegant restraint of your slide deck.

***

Visual synchronization

When presenters speak well but show too much, attention fragments. Visual synchronization means making sure the slide pacing, hierarchy, and transitions reinforce, rather than compete with, your spoken flow. In a well-aligned presentation, every slide appears at the right moment, supports one clear idea, and disappears before it turns into visual noise.

What audiences actually see

In any type of presentation, whether live, hybrid, or recorded, people spend more time looking at slides than faces. That means every visual that you put in front of their eyes either clarifies or competes with what you're saying.

Viewers read at roughly 200 words per minute, while you speak closer to 130. If the slide invites them to read while you're still explaining, they stop listening to you. To prevent that, the slide rhythm must follow the same *Promise* → *Proof* → *Personality* → *Promise Returned* structure as your words. Each visual needs to underline just one beat. Never all four.

Synchronizing voice and visuals

Here's a simple and helpful rule: *A change of slide equals a change of thought*

So, if you're still talking when the next slide comes up, you've either talked too much or you've changed slides too soon. Conversely, if a complex visual lingers while you move on, the audience stays anchored to the last idea.

Slide coordination tips

- ***Promise* slides = clarity**
  Use one headline statement in large type and a single supporting image that hints at intent (a sketch, a place, or a human face). The goal is to orient your audience to your message, not explain it.
- ***Proof* slides = structure**
  Make these slides carry the proof by showing your data, diagrams, or comparisons. Keep the text minimal with labels, not paragraphs. Then narrate what matters rather than reading the slide aloud.
- ***Personality* slides = feeling**
  Favor photographs, process imagery, or one authentic team moment. A candid screenshot of collaboration conveys sincerity more than stock imagery ever can.
- ***Promise Returned* slides = resolution**
  Re-show the opening visual or title, but with one visible evolution. For example a result, a client quote, or a visual transformation. It signals completion on both visual and emotional levels.

Presentation tempo

When the voice tempo of the team and the visual tempo of the slides align, attention stays steady. Here is a practical rhythm pattern for slide decks:

1. *Anchor slide* – title and mood image → 5–7 seconds each

2. *Content slide* – graph, plan, or concept → 25–40 seconds each
3. *Breather slide* – single word or quiet image → 8–10 seconds each
4. *Transition cue* – name or phrase of next speaker → 3–5 seconds each

Building those intervals into your script prevents overcrowding. When you're delivering, always pause one second or one breath longer than feels natural before advancing slides and moving on to your next thought. While it may feel awkward, it comes across as poise, not delay.

In the same way that the visual "white space" we discussed in the last chapter gives the eye a place to rest, these brief intervals give the audience's brain a pause in which they can digest the previous material and get ready for the next.

Learning pace by watching professionals

If you want to refine your own timing, study it in action. Pick a clear speaker, perhaps someone from a corporate keynote, a TED Talk, or a public lecture. (I've always found Steve Jobs' presentations to be a great place to learn compelling delivery and I've also learned a lot from stand-up comedians!)

Don't watch for the speaker's style, but for the *silences.* Count the pauses. Measure the gaps between main ideas and slide changes. Most experienced speakers leave two to three seconds of quiet before a new point, and sometimes as much as five.

Those pauses aren't lost time. They are how the audience catches up and how the presenter keeps control of the rhythm. Notice that the strongest moments often come *after* a key statement, not during it. The speaker looks calm, the slide settles, and only then does the next idea start.

Try replaying a short talk with the sound muted. You'll still sense tempo shifts from the speaker's movements and slide transitions. That's the level of synchronization you're aiming for: a pace that's deliberate, consistent, and confident enough to let ideas, and silence, breathe.

AI and automation aids

There are a number of AI tools that can assist with pacing and visual balance. Use them to test consistency, not to make design decisions. They're useful as checkpoints and a quick way to see whether your deck flows as smoothly as you think it does.

The following list reflects a mix of free-tier and subscription tools. These are widely available commercial platforms as of 2026. Capabilities and privacy policies evolve quickly, so verify functions before integrating them into client-facing workflows.

Purpose	App Examples	Typical output/insights
Visual balance testing	*Beautiful.ai* *Tome*	Adjusts the ratio of text to imagery; flags visually dense slides and suggests simplified layouts
Readability & word count check	*Canva Presenter View*	Highlights slides exceeding a recommended text threshold (≈ 40 words); encourages concise phrasing
Design consistency & hierarchy	*DeckRobot* *Gamma AI*	Evaluates alignment, typography, and color hierarchy; offers auto-formatting to improve uniformity
Slide timing feedback	*PowerPoint Speaker Coach*	Measures speech rate and monitors slide transitions; notifies when visuals or pacing move too quickly

Even light, occasional use of these tools can expose pacing or density issues that you might miss when rehearsing from memory. Think of the insights as diagnostics. They're useful for alignment and refinement, never as creative direction.

Warm Delivery through visual restraint

Most presentation stress comes from the slides themselves. When the visuals are too dense, speakers tend to feel rushed, which shows up in your delivery.

Restraint in the amount of content you put on each slide frees your fluency during delivery. Limit every slide to just one message and remember that, in digital settings, *visual silence* reads as maturity. If you spend ten quiet seconds on a diagram without talking, that pause lets the panel study the graphic and breathe with you. It also projects your composure, which translates as confidence and capability.

Slide rhythm checklist

Before your final run-through:

1. Watch your deck on a shared screen with the sound off. Note where you lose interest as it highlights the slides that need simplification or shorter dwell times.
2. Add a blank, brand-colored "breather" slide after any dense graphic.
3. Mark the verbal cue for each transition where one narrator finishes and the next visual appears.
4. In your script notes, write the intended duration beside each slide (e.g., "28 sec").
5. Do one timed rehearsal where another teammate advances slides for you based on those intended durations, not where you are in your delivery. The lag or overlap reveals any synchronization issues immediately.

Why visual timing matters

I said it earlier, but it deserves to be repeated: Review panels don't only evaluate content. Consciously or subconsciously, they're evaluating composure and clarity under pressure. When your slides, voices, and transitions feel well orchestrated, the whole presentation exudes a calm professionalism. Reviewers will sense your confidence, even if they can't articulate why.

Visual synchronization turns a mere slide deck into an intentional experience. Instead of a collection of visuals that you've cobbled together from previous presentations, it becomes a choreography of pacing, tone, and trust that carries through every screen.

Structure brings consistency, but structure alone doesn't inspire. The best presenters balance discipline with ease, using what we might call *planned spontaneity.* It's the ability to sound fresh and responsive inside a framework you've built with care. Every movement, every pause, every hand-off has been rehearsed, yet the delivery feels unforced. Rehearsal gives you the muscle memory; spontaneity gives you the humanity.

In the next section, we're going to explore how your team can move from analytics-driven preparation to solid human connection in the final performance.

***

Cold Rehearsal/Warm Delivery

To this day I can't understand why so many design professionals—and this seems to apply more to firm Principals than staff— so strongly resist the idea of rehearsing ahead of presentations. I can't count the number of times I've heard, *"I've done this so many times before, I don't need to rehearse."* Or, *"We'll rehearse in the car on the way to the client's office."*

Seasoned performers, with Grammys and Oscars under their belts, still understand the enormous value of practice and rehearsal. Considered one of the greatest pianists of all time, Vladimir Horowitz is often cited as saying, *"If I don't practice for one day, I know it; if I don't practice for two days, the critics know it; if I don't practice for three days, the world knows it."*

If you don't practice, your prospective client knows it, interprets that knowledge as indifference about their project, and selects another firm.

That's why strong presenters treat rehearsal as preparation, not punishment. It steadies delivery, clarifies teamwork, and frees you to sound natural when the stakes are high. Rehearsal builds confidence and the resulting connection builds trust. You need both to hold a panel's attention.

Cold Rehearsal is where you measure and adjust. Warm Delivery is where you trust what you've built and let the story breathe. The difference isn't temperature; it's focus. One looks inward to your timing, slides, and transitions. The other looks outward to your audience.

Cold Rehearsal: The discipline

At this stage, treat your presentation like an experiment. You're testing for error, not emotion.

- **Capture the baseline.**
  Record at least one full run-through in your rehearsal tool or video platform. Note the points of friction where concentration drops, where transitions sag, or where tone hardens.
- **Measure the pattern.**
  Analyze timing per slide, filler-word frequency, and vocal range. If your numbers are tight but your team still feels flat, that's the signal that technique has overtaken intent.
- **Stabilize the structure.**
  Adjust order, transitions, or slide rhythm until the whole team can deliver the story without prompts. Like a stage actor who inadvertently forgets a line, it's only when you can lose your place and recover confidently that the structure is secure enough for Broadway.

Cold Rehearsal creates muscle memory. It moves performance details from conscious thought to instinct, freeing you for the all-important spontaneous connection later.

Planned spontaneity: The bridge between

Every great presenter sounds as though their words are being discovered in the moment. They aren't. They've been rehearsed dozens of times. The difference lies in *planned spontaneity*, which can be thought of as speaking from the memory of *meaning,* not of *wording.*

Here's how teams cultivate it:

1. **Memorize sequence, not script.**
   Know what comes next, but not exactly how you'll phrase it. This keeps speech adaptable and alert rather than recited.

2. **Use your notes as landmarks.**
   But don't try to memorize a script. A single phrase or image cue triggers the next segment; that freedom lets natural language fill in around it.

3. **Expect imperfection.**
   Minor stumbles humanize you. Panels read your ability to recovery as poise, not failure. The only mistake is the panic that follows a small one.

4. **Speak to one person, not to the grid.**
   In hybrid rooms, address a single face or camera frame as though you're explaining an idea directly to that person; it instantly softens tone and invites authenticity.

5. **Welcome the unscripted moment.**
   If a panelist smiles, nods, or interrupts with a question, shift to that opportunity. That one exchange will linger longer in memory than a flawless scripted finish.

Warm Delivery: The connection

On presentation day, nothing new should happen inside your own head. All attention belongs outside it.

- **Check your energy, not your appearance.**
  Before starting, take a deep breath and then slowly let it out. It lowers your pulse and opens your voice. The room feels that composure before you speak.
- **Begin steadily, but not loud.**
  In hybrid settings, microphones amplify confidence more than volume.
- **Watch their faces for cues to comprehension or confusion.**
  Subtle nods, leaning forward, or sudden stillness tell you when to pause, repeat, or clarify.
- **Close quietly.**
  Don't rush to a thank you. Let silence mark the final point before any questions begin. It registers as confidence in your message.

Warm Delivery can't be rehearsed directly. It emerges because the discipline of your Cold Rehearsal removed all the uncertainty. When you trust the structure, your energy redirects from monitoring yourself, to reading the audience.

Two final practice runs

Before presenting to the client or selection panel:

Run #1: Solo playback

Each presenter should video a part of their own section of the presentation. Then immediately replay it once with the screen turned away or the video hidden so you hear only your voice. Listen for pacing, tone, and filler words, and make simple adjustments before the next run.

Run #2: Full team flow

Rehearse the complete sequence as though live. No starting over, no stopping for slide fixes. At the end, debrief with one question: *"Did it sound fresh?"* If it did, you've reached planned spontaneity. Your delivery is authentic and anchored by rhythm you've practiced.

Why this balance matters

Panels sense the difference between enthusiasm and strain. Cold Rehearsal gives you precision; Warm Delivery gives you presence. Together they project what every reviewer is subconsciously looking for—confidence with empathy. Rehearse until you sound prepared, then keep practicing until you sound human.

Now that the spoken elements of your presentation have aligned your energy, wording, pacing, and warmth, our next chapter is going to move us into visual clarity. We're going to explore how the imagery you've created remains after you've left the room and continues to speak for you.

CHAPTER 11
Visuals that clarify, instead of clutter

The visual dilemma

Every firm now produces more imagery than ever before. With renderings, diagrams, animations, fly-throughs, and more, the volume multiplies every year. Yet screens filled with images (beautiful or otherwise) don't always translate into understanding. Paradoxically, as the tools for creating visuals have expanded, attention has fractured and meaning has blurred.

Clients often recall one image, but not what it meant. Reviewers scroll past slides that took days to craft because the story behind them isn't clear. The problem isn't a shortage of visuals; it's a shortage of visual editing.

In the world of architectural and engineering communication, an image can fall into two categories: *illustration* or *explanation.* While illustration can decorate your documents, it takes explanation to advance comprehension. Beautiful color, texture, and atmosphere can draw the eye, but if composition, scale, and labeling don't direct that attention to your argument, even the best rendering becomes wallpaper.

Hybrid selection panels amplify this problem. On a shared screen, a slide that reads crisply in your office can collapse into a tiny rectangle competing with chat bubbles and participant tiles. When that happens, your strong ideas get lost behind glare, type, or animation layers.

So, before you draw one more diagram or export one more rendering, let's ask one vital question: *Does this image clarify the choice you need the reviewer to make?* If it doesn't, it's a distraction, no matter how polished it looks.

Common visual traps

- *Decorative rendering.* Images made to impress rather than to inform. They slow comprehension instead of focusing it.
- *Text overload.* Paragraphs pasted from proposals onto slides. Reviewers stop listening and start reading.
- *Inconsistent visual vocabulary.* Each slide follows a different visual logic with different fonts, scales, and symbols so the audience must keep re-learning how to read your story.
- *Overcrowded slides.* As with your proposal layouts, good visuals need white space where the eye can breathe, and the viewer can take the time to comprehend your message.

These aren't artistic errors, they're cognitive ones. The human brain can hold roughly four to seven visual elements in working memory at a time. Anything beyond that dilutes recall. Simplicity, not excess, is the marker of confidence.

In this chapter you're going to learn how to replace decorative volume with visual clarity. You'll learn to translate your expertise into their immediate understanding. We'll start with the three questions every image must answer. And we'll see how the answers distinguish visuals that persuade from those that merely impress.

Principles of clarity

When every slide competes for the viewer's attention, clarity becomes your only effective form of communication. It's not about stripping work down for minimalism's sake. It's about designing each visual so that the message lands instantly, even across a noisy digital environment.

A simple test applies to every image, diagram, or drawing you present. It should answer at least one of three questions: *Where? How? Why?* If it doesn't answer any of them, it's redundant and you must remove it.

1. Where? To provide orientation for the viewer

This is the viewer's first instinct: *Where am I in this story?* A visual that answers *where* provides both context and scale. Typical examples include:

- Site plans that show recognizable landmarks or orientation cues.
- Diagrams that establish a project's place in a larger system, such as a campus, a watershed, a corporation, or a community.
- Photos that create emotional anchors such as people using space, a view line, and a skyline.

Without answering *where*, your audience can't place themselves inside the idea. They become distant spectators rather than participants. If you add one relatable element, such as a figure, a tree, a vehicle, or a doorway, it establishes scale and helps non-technical viewers interpret physical space instantly.

2. How? To explain the mechanism

The *how* image explains function. It's the unseen diagram that interprets the 'Wow!' shot. It explains how air circulates, daylight moves, people flow, or materials connect. Typical tools include:

- Section perspectives with simple motion arrows or gradients.
- Process diagrams that demonstrate input, transformation, and outcome.
- Phasing graphics that show sequence over time.

The risk here is density. Many *how* graphics fail because they try to show too many steps at once. Limit each to *one layer of logic* per image. Need to explain another layer? Add a separate slide and shift verbally between them. Sequential simplicity always beats simultaneous overload.

3. Why? To uncover the purpose

The most powerful visuals answer *why*, as they translate rationale into emotion. These might be comparison images, post-occupancy photos, or data overlays that reveal impact. They don't just describe outcomes; they reveal *value*. To test each image, ask yourself:

- Does this diagram or image make the benefit unmistakable?
- Can a reviewer see, without narration, what problem has been resolved?

When an image communicates *why*, the conversation changes from "what did they design?" to "shouldn't we build this?"

The clarity filter

Before finalizing any slide or visual, run this quick filter:

Step	Question	Keep or cut?
1	Can a first-time viewer grasp the core message in 3 seconds?	If no, simplify or split it.
2	Does it reinforce your narrative's *Promise* → *Proof* → *Personality* → *Promise Returned* sequence?	If no, re-label or move it.
3	Does it answer *Where, How, or Why* for this section of the story?	If none, delete it.

Clarity isn't created when visuals are first drafted. It happens when they're *edited.* Especially in hybrid settings, every image must perform a big task. It must be fast, clear, and human.

Now that you've distilled what each image must *communicate*, the next challenge is how to generate these visuals efficiently without losing intent. In the next section we'll explore how AI-assisted visualization can translate complexity into clear graphics that still carry your firm's human voice.

***

Generative Design as interpreter

Generative and AI-enhanced tools aren't here to design your projects for you. That's your domain. The architecture, the engineering, the drawings, and the intent already exist. What these tools *can* do is interpret your visuals so that reviewers see what you want them to see, faster and more clearly.

Used well, the technology becomes an *assistant for clarity*: highlighting hierarchies, testing readability, simplifying complexity, and sharpening contrast so your existing images communicate with precision.

Storyboard before creating your slides

The most effective visuals, whether human- or AI-aided, begin with a storyboard. A storyboard is a visual outline that maps how your presentation will unfold. It lays out each slide as a simple thumbnail, clarifying *what* the audience will see, *why* it appears at that point, and *how* one idea connects to the next. You design and refine the storyboard in quick sketch format before you move into detailed graphics or layout. Storyboarding isn't about cinematic polish; it's a thinking structure that aligns image purpose, audience, and sequence before you start refining slides.

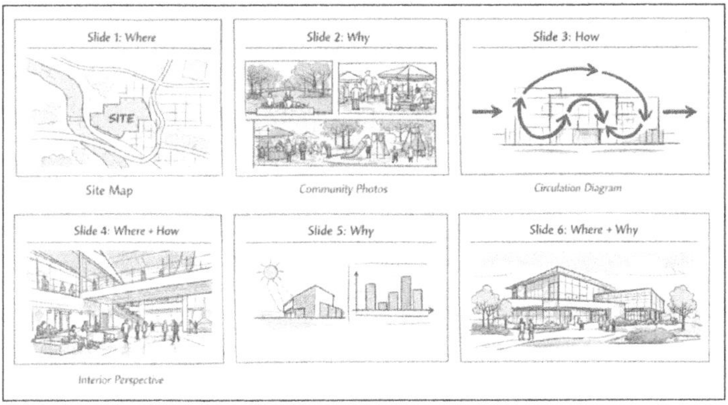

Figure 11.1 (previous page): Sample storyboard. Clarifying *"Where, How, and Why." A storyboard translates a presentation outline into a sequence of visual intentions. Each thumbnail defines what the audience should see and understand at each stage of orientation (Where), mechanism (How), and purpose (Why) so teams can adjust order, emphasis, and pacing before building final slides.*

Follow these steps to create your pre-slides storyboard:

1. **Define the goal for each frame**

   Label every slide or image with what it needs to accomplish:
   - *Where* does this orient the viewer?
   - *How* does it explain process?
   - *Why* does it matter?

   This step decides the function of the slide before you create its form.

2. **Select your existing materials**

   Use the renderings, plans, photos, or diagrams from your project library that you would normally rely on to explain your approach. It's rare that anything new has to be created, since you've got the images of your previous projects and you've likely created the design sketches for any new ones. The storyboard lets you organize the material into an intentional flow.

3. **Note the audience's vantage point**

   On the storyboard, sketch or list whether each image is intended to be seen on a large screen, a laptop, or a mobile window. That context determines how much detail can survive.

4. **Identify what needs clarification**

   Mark where information is likely to get lost such as text being too small, colors blending, contrasts too subtle, or multiple focal points competing. These marks cue the AI-aided refinements you'll run later. A half-hour spent storyboarding will save you hours of slide-level correction.

Using AI to clarify what you already have

Once your storyboard defines the purpose you've set for your presentation, AI becomes a diagnostic and editing layer, not a design engine.

1. *Annotation and overlay aids* Use image-editing AI (such as Adobe Firefly, Canva Magic Design, or Gamma AI) to add clean overlays such as arrows, flow lines, or zone highlights to drawings that already exist. Prompt for transparency and consistency so your work stays visible and clear beneath the annotation.
2. *Contrast and focus testing* Upload draft slides to visual-clarity testers or presentation assistants (e.g., DeckRobot, PowerPoint Designer, Microsoft Speaker Coach). They'll flag any issues with text-to-background contrast, type hierarchy, or overcrowded frames that would fatigue viewers.
3. *Audience-ready clarity checks* AI tools can act as fresh eyes on drawings that designers already know too well to see clearly. Upload a slide or rendering for interpretive feedback rather than generative output. Many platforms now simulate how different audiences perceive complexity, including testing whether key routes, focal points, or daylight patterns are immediately understandable. For example, you might prompt: *"Identify the primary elements a first-time viewer will notice in this plan and suggest how to emphasize the circulation path without redrawing."* The goal isn't to simplify the design but to **clarify its communication** and ensure that the visual hierarchy supports comprehension, not just composition.

AI refinement pass

1. Storyboard your slide sequence.

2. Run one image from each "Where/How/Why" set through an AI clarity or accessibility check.
3. Compare before-and-after versions with your team.
4. Keep only what improves readability and narrative flow.
5. Repeat the pass once more before final delivery. The process often trims 10-15 percent of unnecessary visuals while strengthening the remaining ones. No client will ever criticize you for showing less instead of more.

Why the interpreter mindset matters

AI can multiply images infinitely. But design communication thrives on reduction. What moves clients is clarity born of judgment, the sense that a real person has chosen what to show and what to leave out.

When you use the tools to condense, emphasize, or test, rather than to invent, you position technology as a translator of your thinking, not a substitute for it. Clear visuals show mastery. Over-decoration looks like insecurity.

Now that you've built and refined your storyboard for clarity, the next step is to establish and maintain consistency. In the next section I'm going to show you how to create a coherent visual language, using color, contrast, typography, and hierarchy, that keeps every screen readable, and every reviewer included.

***

Visual grammar and accessibility

Just as verbal grammar helps a sentence flow, visual grammar guides the eye. Imagine if someone switched mid-conversation from the polished cadence of an Oxford lecture to the quick slang of a Brooklyn street corner. The meaning might still be there, but the shift would jar the listener.

Visual grammar is the consistent system your slides 'speak' with. It coordinates color, hierarchy, typography, and contrast to help every viewer, including those with visual differences, process your ideas without friction. If your visual grammar is inconsistent, it pulls the viewer's attention away from your content to focus on your style. (Or lack of it.) While your intent might be clear and your images strong, you can still lose your audience if that visual grammar breaks mid-sentence.

Your clarity can earn the client's trust and your consistency will earn their comprehension.

1. Color should communicate, not decorate

Color should carry meaning, not mood. Too often, color palettes borrowed from marketing campaigns or visualization filters blur technical accuracy. Restraint should always be your guide. Just one highlight color, supported by neutrals and accessible contrast.

- *Assign purpose to color.* Cool tones might represent context, warm tones signal human activity, and neutral grays hold background elements.
- *Test accessibility.* Ensure enough contrast between text and background so it remains legible to everyone, including readers with low vision or color-blindness. You can check this using any of the free WCAG* contrast-testing tools. The common benchmark is a 4.5:1 ratio between text and background.
- *Maintain consistency.* A directional arrow that's blue in one slide shouldn't turn green later. Reviewers subconsciously rely on repetition to track logic.

Guideline: If a color isn't telling the viewer what to notice or how to understand, remove it.

2. Typeface as tone

Type is your spoken voice on screen. If it's too ornate, it distracts. If it's too faint, it disappears.

- *Use one family, two weights.* A clear sans serif like Helvetica Neue, Arial, or Calibri works across screens. One weight for body text, one for emphasis.
- *Avoid mixed personalities.* Combining modern and serifed styles fractures hierarchy quickly.
- *Check scale.* For shared-screen presentations, no body text smaller than 24 points. Hybrid reviewers often join from phones or scaled windows.

When in doubt, treat every line of text as its own micro-diagram with just one idea and one visual priority.

3. Hierarchy and flow

Viewers need to know where to look first, second, and last. That path is your slide's syntax.

- *Anchor each slide with a single focal point.* Whether that's a plan, a phrase, or a number, let all the other elements support it.
- *Use spacing as punctuation.* White space separates the points you're making in your visual argument. Embrace it, rather than fear it.
- *Repeat key positions on the slide.* If titles and labels shift unpredictably, audiences spend cognitive energy re-learning the layout rather than absorbing the message.

A consistent structure allows reviewers to predict where information will appear. That comfort accelerates their comprehension.

4. Accessibility by design

Making visuals accessible isn't just compliance, it's respect for every participant in the conversation.

- *Contrast and legibility.* Prioritize clarity over subtlety. While muted tones may read well on a calibrated display, they're likely to vanish on low-brightness laptops.
- *Alt text and verbalization.* If a reviewer can't see an image, make sure your script (or AI presentation tool) includes verbal

equivalents: *"This diagram shows airflow moving left to right through the filtration system."*

- *Motion sensitivity.* Avoid animations during live reviews. They raise cognitive load and can even cause motion sickness in some people.
- *Color-independent explanations.* If a diagram relies on color to distinguish zones, add labels or patterns so it remains legible to color-blind viewers.

Your goal is to make your presentation universally accessible, using visual language that communicates the same message, regardless of how or by whom it's seen.

The readability audit

Before any major review, run your deck through a quick readability audit:

1. View the slides in grayscale and miniature thumbnail mode.
   If meaning vanishes, color or scale is carrying too much weight.
2. Test for blur.
   Step back six feet or reduce view to 25%. Can you still grasp hierarchy at a glance?
3. Read aloud.
   If you have to explain every visual element, it isn't doing its job.
   Small consistency checks prevent major credibility leaks.

Why it matters

A coherent visual language silently communicates authority. It signals that your mind is as organized as your work. While reviewers won't consciously note your typography or color balance, they'll simply find your story easier to comprehend, and therefore to trust.

When your slides read cleanly to every eye and every screen, you've achieved the visual equivalent of good conversation. There's no noise, and no strain, just meaning.

With visual grammar and accessibility established, our final step is presentation rhythm. In the next section we're going to study how to pace, pair, and speak to visuals so they reinforce delivery without splitting attention.

** Note: "WCAG" stands for Web Content Accessibility Guidelines, the international standard that defines readable contrast levels for text and images across digital media. You don't need to memorize the numbers, just aim for clarity on every screen.*

From image to insight

Even when a visual looks complete, clarity only emerges through review. Strong teams treat image development as an iterative dialogue between what the presenter *intends* and what the viewer *actually comprehends.*

The difference between an adequate slide and a persuasive one usually isn't another rendering pass. It's a structured feedback loop that turns image into insight. This loop doesn't demand new software. It demands a steady habit of *testing, interpreting, and refining.*

1. **The three-step cycle**

   Every refinement session should follow three repeating passes:

   Step 1: Generate drafts (fast and frequent)

   Create the first pass of your visuals loosely. Their purpose isn't approval, it's observation. Invite people who haven't been involved in the process to describe, in one sentence, what they see *before* you explain it.

   Their answer reveals whether the *"Where, How, or Why"* message is landing. If their interpretation matches your intent, the graphic works. If it doesn't, change the visual before moving on to polish the style.

Step 2: Curate by people for accuracy and tone
Use the judgment of real people, not AI output, to decide what the slide is truly *saying.* Ask yourself: *Is this visual helping the audience understand a choice, or just impressing them with polish?* Keep only those visuals that advance your argument or your client's confidence.
The Five-Second Read
Share the image with a colleague unfamiliar with the project. If they can tell you its main idea in under five seconds, it's communicative. If they start guessing, it's illustrative, not explanatory.

Step 3: Refine the captions to guide interpretation
Never assume that a visual will always speak for itself. A good caption can transform a graphic into guided insight. It will frame what the audience should notice, not what the designer assumed they will.
A strong caption links purpose to evidence: *"Energy model results showing passive shading's 30 % daylight gain."*
Avoid narrating the obvious *("Site Plan of Phase Two")*, and instead anchor understanding. Each caption should clarify *why* the image matters in the story, not restate *what* it shows.

2. AI as mirror, not editor

AI can accelerate testing, but it shouldn't dictate revisions. Use visualization-support tools to uncover where the slide might confuse, not to redesign your work. Examples include:

- *Clarity tests* (e.g., DeckRobot, Beautiful.ai, Gamma AI) can flag overly dense slides or flagging contrast.

- *Accessibility scans* reveal color-blind contrast issues or type sizes unreadable on mobile screens.
- *Image-summary models* can output a one-line "machine caption." Compare that auto-caption to your intent. If the AI's description misses the point, the audience probably will, too.

Treat this feedback as a *mirror of perception*, not a creative directive. AI can supply the outside view, but people decide the meaning.

3. Turning critique into consensus

Structuring the review loop helps teams speak a shared language about images. Here's a quick framework for critique sessions:

Reviewer Prompt	Why It Works
"What do you notice first?"	Tests hierarchy and focus.
"What question does this image answer?"	Confirms its *Where-How-Why* role.
"What might confuse a non-designer?"	Surfaces jargon or overly subtle contrasts.
"Does the caption match what you see?"	Tests alignment of text and image meaning.

Always keep feedback factual, not personal. Replace *"I don't like it"* with *"I didn't understand it until you explained."* Clarity grows when critique measures comprehension, not personal preference.

4. From visual to verbal insight

A refined image has two outcomes:

- It's *easy to follow* and the viewer grasps the message quickly.
- It's *faster to discuss* because it tells the story clearly.

The ultimate test isn't whether the graphic looks polished, but whether it earns a comment in the meeting that drives insight: *"That's the diagram that made me realize how circulation solved the congestion problem."* Once an image can provoke a sentence like that, it has fulfilled its purpose.

Why this loop matters

Without a structured review, firms tend to overproduce images rather than clarify the ones that are needed. A disciplined loop stops that spiral. By generating quickly, curating critically, and refining captions deliberately, you preserve the human voice inside the technical imagery.

The result isn't more slides. It's more signal and less noise. When feedback loops convert visuals into shared understanding, your presentation shifts from *showing work* to *showing wisdom.*

With a reliable review loop in place, you can now evaluate your visuals through the audience's eyes: how fast they comprehend, how confidently they interpret, and how they feel while viewing. That's what we're going to explore in the next section.

***

The Empathy Test

Even the most elegant slide has no value if its audience feels left out of it. *The Empathy Test* asks one question that transcends design and technology: *Do your visuals feel clear and considerate to someone who wasn't in the room when you made them?*

The purpose isn't sentimentality. It's accessibility for people-to-people connection. It's building imagery that welcomes every reviewer, regardless of background, attention span, or screen size.

1. Understand the audience's starting point

Every reviewer arrives with a different mental map. Some read drawings fluently, while others see only color and shape. Empathy begins by asking, *What can they know in three seconds, and what do they need me to show them next?*

Try this quick calibration exercise:

- *The analysts* scan for logic, so data sequences should be shown first
- *The community representatives* look for people, so include human scale early
- *The clients or investors* seek reassurance, so begin with purpose, not process

Designing empathy means starting from the receiver's perspective, not your own production order.

2. Model how they will experience the review

As best you can, try to replicate the review environment your audience will actually face.

- *Shrink the screen.* View your slides on a phone or small laptop to see where readability breaks down.
- *Add distraction.* Keep chat windows open or play quiet background noise; note where focus slips.
- *Simulate fatigue.* Run through your deck after a full workday. If you find yourself tuning out, so will they.

Empathy is not a design flourish. It's testing the conditions of real-world attention. If an idea survives under stress, it will shine in calm.

3. Translate complexity into comfort

As I've repeatedly emphasized, we're going for both clarity and empathy. But these can collide when we show technical depth. The goal isn't to simplify *content*, or 'dumb it down.' It's to simplify the reviewer's ability to quickly grasp the *what and why* of your message. If you suspect your point might get lost in the details, try these empathetic adjustments:

- *Front-load meaning and hide the mechanics.* Headlines should reveal purpose, with technical data tucked beneath.
- *Pair visuals with plain-language cues.* A small phrase such as "How clean water gets delivered to the development" gives everyone a foothold.
- *Show people where people matter.* Include a figure, a user, or a moment of activity. This will reveal scale and intention faster than measurement ever can.

If a viewer feels smart while reading your slide, they will trust you more.

4. Invite reaction, instead of evaluation

Empathy in your presentation isn't only about what you show, it's about how you listen. The goal isn't to defend your design decisions, but to understand how your audience experiences them.

After you finish a rehearsal, invite one person to watch your deck silently, without explanation. Then ask: *"What stayed with you?"*

Encourage honesty. Did the slides leave them calm, curious, confused, or overloaded? Their emotional temperature matters just as much as their technical understanding. Often the comments that you least expect *"I wasn't sure what I was supposed to look at,"* or *"That part made me relax"* will teach you more than any formal critique.

During your team review, try asking open prompts such as:

- *What emotion did this slide bring up first?*
- *Where did your attention drift?*
- *Which moment felt obvious, even without explanation?*

These questions keep the focus on experience, not justification. They shift feedback from *"Do you like it?"* to *"Did it reach you?"* And that's the heart of empathy. When a reviewer feels acknowledged rather than targeted, understanding follows naturally. The conversation moves from correction toward connection.

5. Measure empathy as inclusion

The final proof of empathy is inclusion. Your presentation should carry the same meaning to every viewer, no matter their device, eyesight, or environment. Simple checks can make a big difference:

- *View your slides in grayscale.* If hierarchy disappears, color is doing too much work.
- *Play it silent.* Mute your narration and see if the story still makes sense on visuals alone.
- *Test captions and transcripts.* In hybrid settings, captions aren't extra, they equalize access. Even if connection quality wobble, they ensure that everyone can follow your thought.

Inclusion isn't a compliance task. It shows that you care about who's on the other side of the screen. When audiences realize you've thought about *their* comfort, their trust deepens before you even speak. Clarity becomes empathy when every participant can see, hear, and interpret the same story with ease.

Why all this focus on empathy matters

Where your confidence will earn their attention, your empathy will earn their belief. While your technical precision lets people know that you're capable, your empathetic design tells them you care.

When a reviewer feels guided rather than overwhelmed, your work reads as thoughtful, not theatrical. By designing your presentation with empathy, you replace the old rule of *knowing your audience* with a better one: *sharing their view.*

That's when a presentation stops feeling like a performance and starts feeling like a dialogue. It becomes an exchange of understanding instead of a display of expertise. Every stakeholder who attends or watches your presentation, regardless of where they're coming from, can find themselves in the story you're telling.

The 3-second empathy check

Before your next rehearsal, try this quick exercise on a single slide. It takes less than a minute and offers instant perspective.

1. *Flash it briefly.* Show the slide full-screen for just three seconds, then hide it. No setup, no commentary.
2. *Ask one question.* "What stayed with you as a feeling or understanding?"
3. *Listen to their first reply.* That's where perception lives.
    - If they describe your *idea*, your hierarchy is clear.
    - If they describe only *style or color*, polish is outpacing meaning.
    - If they remember little or nothing, simplify and retest.
4. *Adjust once, not endlessly.* One informed revision does more for empathy than ten rounds of decoration.

When the meaning lingers after three seconds, connection has begun.

Case Study: Clarity recalled

A midsize engineering firm was preparing its annual sustainability submission for a municipal infrastructure client. The brief required a statement of *"operational energy impact reduction targets."*

Their first draft filled an entire page: 300 words describing pump efficiencies, control sequences, and load-balancing algorithms. It was technically flawless and instantly forgettable. When asked, reviewers said they *"skimmed and moved on." It was a classic case of TL;DR (Too Long; Didn't Read)*

In the second round, the team reframed the story visually. Instead of paragraphs, they built a single layered diagram. At the center sat one bold circle labeled *"100 % System Load."* Radiating out from it, translucent rings illustrated incremental reductions:

Flow control (-18 %)

Pump optimization (-9 %)

Daylight-linked ventilation (-7 %)

Heat recovery (-14 %)

A simple blue gradient tied every layer back to water movement, the project's core system. They added a title and a subtitle across the top.

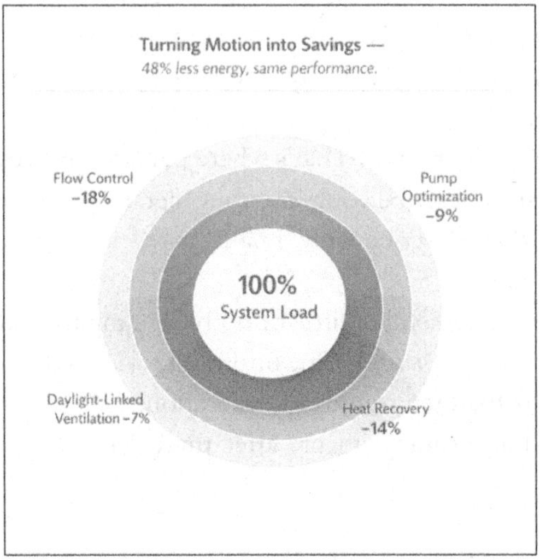

Figure. 11.2: *Turning Motion into Savings. Simplified performance diagram showing how integrated control strategies reduce total system load by 48 percent while maintaining full performance. Energy savings are distributed across four measures — flow control (–18 %), heat recovery (–14 %), pump optimization (–9 %), and daylight-linked ventilation (–7 %) — illustrating the power of clear, visual storytelling in proposal and presentation graphics.*

During the presentation, no one read a word of the original narrative. But every reviewer recalled the image days later. In follow-up interviews, two panelists said they could still *"see the rings of savings"* when they talked about the presentation.

By converting technical detail into visual proportion, the team didn't dilute the data, they gave it meaning at a glance. The diagram told a complete story with measurable results, mechanical logic, and human benefit, united in one composition.

Every slide you create is more than a picture, it's a moment of translation. It carries your thinking across distance, discipline, and distraction. When you remove clutter, build empathy, and design for comprehension, your visuals stop competing for attention and start building trust. That's what we've been trying to achieve in this chapter: visuals that don't decorate ideas, but *deliver* them.

You've moved from the principles of clear composition (*Where, How, Why*), to consistent language, to iterative refinement, and finally to empathy, the invisible skill that lets your audience see themselves inside your message.

But, up to this point, our focus has been on the individual presentation. We've concentrated on what one project team can do to convey meaning faster and make it easier to grasp. But clarity doesn't stop with a well-edited set of slides. To last, it must extend across teams, projects, and tools. It must become a shared visual language that makes every touch point from your firm instantly recognizable and reliably understood. Clarity begins to represent the culture of your entire firm.

In the next chapter we're going to step back and explore how your firms can extend this clarity across teams and tools, so every presentation shares the same visual DNA. We'll learn how you can create templates and visual systems that guide every project team without stifling creativity and build a framework that keeps your firm's story consistent, no matter who's telling it or where it's shown.

PART 4

LEADING TEAMS THROUGH THE CHANGE

CHAPTER 12
Digital Confidence for Non-Tech People

Confidence before competence

Most design professionals never set out to be "tech people." They set out to solve problems, win work, and serve clients well. The technology just showed up. For some it came early in their careers. For others, it arrived mid-career. And for many it came just as they had begun to contemplate retirement and wondering whether it was worthwhile to learn yet another new skill. For everyone, it began asking for attention it hadn't earned yet.

If you've ever looked at a new AI tool and felt that flicker of hesitation—*Do I need to understand what's under the hood before I touch this?*—you're in the majority. The truth is, digital confidence isn't about knowing the machinery. It's about knowing that your judgment still matters even when a new system enters the room.

When someone on your team says, *"I'm not technical,"* what they usually mean is, *"I'm not sure I'll get this right the first time."* It might even mean, *"I don't think I have the patience or bandwidth to learn yet another new thing."*

That's not ignorance. That's professionalism and human nature coming together to protect you. We've spent careers being rewarded for knowledge and precision, not experimentation. And many of us have already absorbed the integration of drastically new technologies: CAD, BIM, Lidar, CRM… The list seems endless.

And now, AI has arrived and flipped the script again. In this case, it's inviting people to *try things before they're certain.* For many in the world of the built-environment, that's way outside the comfort zone. But it's also entirely learnable.

So, let's set a new baseline:

Digital confidence means you can explore without breaking anything that matters. It's the assurance that you can open a tool, test a prompt, and close it again. Still safe, but smarter than you were five minutes ago.

Early experiments don't require mastery. They require *permission*. Permission to play a little. To explore AI in low-stakes moments like drafting a summary paragraph, re-wording a proposal bullet, or sketching a headline. In those moments, you're not training software, you're safely expanding your comfort zone.

It's like trading your wired earbuds for wireless ones. The music hasn't changed, but the way you connect to it has. At first you fumble for the pairing button, but a week later you forget you ever had a cable. What once felt uncertain becomes effortless rhythm.

Every firm that thrives in this transformation will have leaders who normalize that kind of curiosity. When people at every level can say, *"Let's see what the machine comes up with,"* without fear of sounding naïve, that's when innovation starts to feel safe.

I'm reminded of a story I once heard of a guy who was reluctant to dance. He kept telling himself that he'd try it after he become more confident. And how do you become more confident as a dancer? By dancing. Confidence doesn't come after competence. It's what allows competence to begin.

Permission to play: How to experiment safely

If confidence lets you open the door, curiosity keeps it open. But unstructured curiosity can feel like chaos, especially inside busy project offices where every hour must be justified. That's why the smartest firms treat digital exploration more like a safety drill than a gamble. You plan it, label it, and de-risk it.

The first rule of safe exploration:
Keep the stakes microscopic.

No one learns to swim by diving off the high board. They start in the shallow end where you can't get into trouble. And they dip their toe in for just five or ten minutes at a time. In the same way, your first AI experiments should live in low-impact spaces. You start with a draft paragraph, an internal summary, a simple email rewrite. Work that won't go directly to a client until a person decides it's ready.

Begin with curiosity, not challenge. Ask a system something ordinary:

- *Summarize this page for me in plain language*
- *Give me five alternative headlines*
- *List the three points my team should discuss in our next kickoff*

These are small tests of function, not faith. You're learning how the tool thinks, not betting on its brilliance.

The second rule of safe exploration:
Experiment out loud.

Try it in pairs or small groups so discovery feels shared. When one person explores and another one watches, both learn faster. You stop being "the person taking a risk" and become "the team running a test." That small social shift turns vulnerability into curiosity, and curiosity into capability.

The third rule of safe exploration:
Write down your wins, however small.

That first time the system gives you a result that makes sense and will work for you, circle it. Celebrate it. When people see visible proof of saved effort or sharper clarity, skepticism softens. Momentum starts with micro-successes.

I guarantee that some of your early results will make you laugh or shake your head. The hallucinations that can be generated as you find your way around it can be very entertaining. So, recognize that you're only putting your toe in the water. Think of this phase as playing in a sandbox, not preparing a showcase. You're not trying to prove anything, right or wrong, about AI. You're just experimenting for yourself and your colleagues. And you're learning that these tools can work *for* you, not *instead of* you. By treating early use like exploration instead of expectation, you reframe the entire learning curve. It's not a test of intelligence. It's an act of professional curiosity.

If confidence opened the door, and curiosity kept it open, permission to play lets you explore what's inside.

Pairing people and prompts: The questions that shape better output

Once you've given yourself permission to experiment, the next challenge isn't learning new technology. It's learning a new way to *ask*.

In the world of design and communication, we've long been taught that answers matter most. In the world of AI, the quality of your questions decides everything.

Prompts are simply questions in disguise, and they need the same craft we bring to client interviews or design charrettes. A good prompt is part direction and part context. It tells the machine what you want, while teaching it who you are.

Here's the simplest structure to start with:

Context + Action + Outcome

Example:

"You are helping a mid-size architecture firm refine a project summary for a sustainability award submission. Rewrite this paragraph for clarity and tone so it feels professional and optimistic."

That's the difference between vague automation and guided collaboration. You've explained the scene ("who"), the task ("what"), and the intent ("how it should feel"). The system now has something to work *with* instead of just something to work *on*.

Whenever you prompt, you're not handing off your thinking, you're framing it. Think of it like briefing a skilled intern. If you give them only a file and a deadline, they'll flounder; if you add purpose and tone, they'll flourish.

Prompts also work best as conversations, not commands. Instead of chasing the perfect one-liner, start broad and refine. Try three quick rounds:

1. *Draft* – ask for a first take.
2. *Adjust* – tell it what you like and what's missing.
3. *Direct* – ask for that one missing element with clarity.

This rhythm of ask, adjust, and direct builds familiarity fast. Within a week, you'll start sensing what kind of wording unlocks the improved results. It's not "learning to code," it's "learning to coach."

And don't forget tone. Words like "*clear*," "*trusted*," "*human*," and "*succinct*" are magic to these systems. They represent intent, not jargon. They remind the machine that your goal is understanding, not polish for polish's sake.

Try experimenting with these safe starter prompts for daily work:

- *Simplify this paragraph for a non-technical client without patronizing or losing professionalism.*
- *Draft a friendly, engaging, but enticing email that invites a client to review our concept sketches.*
- *Turn these bullet points into a three-minute meeting summary.*
- *Explain these technical steps in plain English for a community group.*

Each prompt builds more comfort with the language of guidance. Then try some of your own, using your own language. Don't be afraid to ask a wrong question. The more interaction you have with the AI tool, the more you'll learn how it responds and the more it will learn about you and the context in which you're working.

And just like any good conversation, *continuity matters.* If you start over with a brand-new chat or thread, the relationship resets, and the system forgets everything you've already established including your context, your tone, and even the examples it mirrored back well.

Keeping a thread open as you refine ideas allows the tool to build a richer sense of your direction. You don't have to complete a conversation or interaction in one sitting. You can leave and come back to it later. Just make sure you resume the chat in the same thread. End the thread only when that conversation, interaction, or learning cycle is complete.

For example, in the writing of this book, I've been interacting with ChatGPT in the same, ongoing thread for several months now. We've had conversations that establish the voice and tone I want to write in, the overall table of contents, the pacing and word count I want to maintain, and we're keeping track of the various ideas that have sprung up along the way that I intend to incorporate before we're finished. If, at any time, I had lost this thread, or started a new one, all the learning that this assistant had achieved and all the references we've cited would be lost and I'd have to begin the training and orientation process all over again.

As you interact more and more with AI, you'll also begin to notice that the act of writing prompts sharpens how you interact with real teammates. Learning to clarify what you want before you ask for it, whether you're asking a machine or a person, creates better communication everywhere. When that happens, AI stops feeling like a robot and starts feeling like a rehearsal space for your own clarity.

Psychological safety: How leaders protect experimentation

Curiosity never survives in an atmosphere of anxiety. It survives in places where people can say, *"I tried something new"* without scanning the room to see who's judging.

Every firm says it wants innovation, but innovation only takes root where small failures are treated as tuition, not as trouble. In Chapter 3 we talked about how the firms that improve fastest are the ones that fail fast and fail forward. In the early stages of AI adoption, that mindset is everything.

Leaders don't need to be (and perhaps, shouldn't be) the first to master every tool. They just need to model visible *comfort with not knowing*. When a principal or manager says, *"Show me what you discovered,"* instead of *"Why did you spend time on that?,"* it changes the culture immediately. Curiosity gets a hall pass.

A few practices make that safety real:

1. **Rename "mistakes" as "discoveries."**
   When an experiment goes sideways with wrong output, strange wording, or an "hallucination" that makes everyone laugh, log it. Treat it as a note, not a knock. This reframes imperfection as data instead of disappointment.

2. **Draw the care lines early.**
   Make clear what *is* and *isn't* appropriate for experimentation. Internal drafts? Free game. Client-facing deliverables? Human review required. People feel emboldened when the boundaries are visible.

3. **Create small, recurring show-and-tell moments.**
   A five-minute "What I Tried This Week" segment during the staff meeting or over coffee does wonders. It normalizes effort, not excellence. Over time those tiny shares compound into organizational intelligence.

4. **Reward curiosity publicly.**
   Praise the behavior, not just the product. Acknowledge someone for *asking a smart question* or *testing a safer workflow,*

even if the output itself was forgettable. Recognition cements safety faster than any policy memo.

Leaders can also serve as translators between excitement and anxiety. Many team members watch AI advances with mixed emotions. They want to keep up, but also to keep control. By speaking frankly about your own learning curve ("I'm figuring this out alongside you"), you create permission at scale.

One practical move is to pair seasoned professionals with digital explorers on a short-term task. The experienced partner will anchor the necessary judgment while the explorer brings technical agility. Nobody loses dignity, everyone gains perspective, and the firm gains blended competence.

Psychological safety doesn't come from an inspirational poster on the wall. It's a pattern of small reactions that tell your people, *"It's safe for you to learn here."* Once that confidence is internalized, experimentation no longer feels risky. It feels like routine growth.

When leaders protect that comfort, the curiosity you sparked in individuals starts to multiply. Teams begin to share discoveries instead of hiding them. AI stops being *"a tool in the corner that's going to replace me"* and starts becoming part of your firm's collective intelligence.

Measuring small wins: Turn confidence into momentum

The fastest way to make a big change stick is to make it *visible.* Confidence becomes culture only when people can see and measure its results, however small.

In the early stages of the adoption of any new technology, progress doesn't show up in project margins or grand metrics. For those who lived through them, the early days of CAD and BIM integration were downright painful!

It's the same now with the adoption of AI. Progress shows up in minutes saved, in friction reduced, and in the way people talk about their work. If you can spot and name those micro-wins, you multiply them. Start with **simple, human measures**, the kind that would fit on a Post-it or a coffee-cup sleeve:

- *Time reclaimed.* "That outline took me 10 minutes instead of an hour."
- *Noise reduced.* "It cleaned up and streamlined a draft report that had me stuck."
- *Insight surfaced.* "It pointed out something we hadn't noticed."
- *Energy preserved.* "It helped me start—so I actually finished."

None of these will headline your annual report, but together they prove the momentum is real. They demonstrate that the technology is earning its place instead of demanding it.

Encourage teams to track micro-metrics over perfection metrics. Maybe you keep a shared *"Quick Wins"* page on your intranet, or a whiteboard that collects weekly victories: *"AI cut my rewrite time in half," "Got three headline options from one prompt," "Found a typo we missed."* Each highlight becomes a behavioral cue and a reminder that trying is worth it.

Quantifying early progress also builds credibility with skeptics. When leaders can say, *"We've piloted six micro-tasks, and four saved over 30 minutes each,"* the conversation shifts from *belief* to *evidence.*

If you prefer structure, think in terms of *"Learn-Apply-Share"* loops:

1. *Learn* – Test a tool in a low-stakes task.
2. *Apply* – Use what worked again on something slightly harder.
3. *Share* – Tell someone what you noticed so the insight doesn't vanish.

These three verbs keep momentum cycling. The loop is small enough to run weekly and concrete enough to show progress.

The truth is that sustained innovation rarely begins with grand plans. It begins with the satisfaction of seeing something get easier, and then getting excited about doing it again. When a single success is noticed, recorded, and repeated, it evolves into process improvement. When a dozen of those tiny stories accumulates, you've just rewired culture without a single mandate.

So, measure what matters most at this stage: *confidence itself.* Notice how people talk about the work. Watch for the shift from *"Can we?"* to *"Let's see what happens if…"* That's the sound of momentum taking hold.

Try this now: Your first 15-minute AI experiment

Now let's make this real. Before you click or flip to the next chapter, carve out fifteen minutes. Find time with no meeting, no pressure, and no deadline to try one small, practical experiment. The purpose isn't to produce a masterpiece. It's to *watch yourself learn.*

Step 1: Pick a Problem That Doesn't Matter Much

Choose something you'd normally handle by hand but secretly wish took less time.

Examples:

- Cleaning up an internal email so it reads more clearly
- Drafting an opening paragraph for a proposal section
- Condensing yesterday's meeting notes into bullet points

Keep it light. You're testing process, not performance.

Step 2: Frame it as a prompt

Write a request that follows your new structure: **Context + Action + Outcome.**

"You are a gifted wordsmith helping me simplify this project summary for our internal report. Rewrite it so it's concise, friendly, and clear to a non-technical reader."

Hit *"Enter,"* skim the result, and don't judge it. Instead, coach it. Tell the AI what works, what doesn't, and what tone you want fixed. Run two or three quick cycles of *Ask, Adjust, Direct.*

"I really liked how you summarized the section on community feedback so keep that up. But could you expand the discussion on the new library parking lot so neighboring property owners will feel heard?"

Step 3: Capture what changed

After your third iteration, jot down two notes:

- *What improved most?* (Speed, clarity, organization, tone?)
- *What still required your input?* (Judgment, nuance, authenticity?)

These answers are your real metrics. They tell you where AI supports your workflow and where your human value shines brightest.

Step 4: Share one thing you learned

Send a short note or post on your team chat: *"Tried a 10-minute AI rewrite of Tuesday's summary—saved half the time and kept the voice."*

That ten-second share normalizes exploration and gives someone else permission to try.

Step 5: Leave the door open

Don't close the thread. Keep the same chat active so the system "remembers" what you're working on. The next time you open it, pick up exactly where you left off. That continuity is how small play can become a reliable partnership.

Confidence opened the door. Permission to play let you step through it. Now you've taken your first walk around the room.

In fifteen minutes, you'll have done more than "experiment with AI." You'll have proven to yourself and to your team that learning keeps pace with curiosity, and confidence is a skill you can build on demand.

What begins as curiosity with a new tool soon becomes collaboration with a new kind of teammate. Once you feel comfortable prompting, refining, and measuring small wins, the next logical question emerges: "*If the system can assist me this well, what happens when it starts assisting the entire team?*"

That's where the shift from *tools to agents* begins. From systems that merely respond, to systems that anticipate. You'll see software that remembers patterns, assigns follow-up tasks, or schedules next steps without being asked. In that new environment, your digital confidence will anchor everyone else's composure. Because confidence scales the same way technology does: by trust, by clarity, and by shared purpose.

In the next chapter, we're going to look at how *assistants* cross the line into *agents*. We're going to explore how you can decide which responsibilities to delegate, what oversight to maintain, and how to build a digital partnership that amplifies your good judgment rather than replacing it.

CHAPTER 13

Agents and Assistants : The next wave of AI

From tools to teammates

If you've made it this far, you've already done more than most. You've dabbled, tested a few prompts, maybe drafted a paragraph, or summarized a meeting. You've seen flashes of how AI can lighten the load without hijacking your process.

That's the foothold you need. Because what comes next isn't about learning more tools, it's about meeting a *new kind* of helper.

Up to now, everything you've used has been transactional. You open a window, ask a question, get a reply, and close the tab. Useful? Absolutely. But those short-term tools forget you as soon as you log out. They can't carry knowledge forward from one project to the next, or one team member to another.

The next wave of digital support looks different. It doesn't just respond, it *remembers*. These are the systems increasingly referred to as *agents* or *assistants*. They are small background programs that persist between sessions and help maintain continuity. Think of them as digital coworkers who don't need constant supervision, but still work under your direction.

They're showing up across industries already. They include tools that check proposal deadlines automatically, compile lessons learned from project folders, or remind you when a client follow-up is coming up. They don't think like humans, but they do keep track, which makes them incredibly useful.

If the AI tools you've used so far are like calculators—quick, one-task utilities—these assistants are more like spreadsheets.

They store context, connect information, and keep working between sessions.

Once a firm has a little momentum with digital confidence that you've built with the habit of safe experimentation, this becomes the next logical step. You'll go from asking, "Can AI help me write this paragraph?" to "Could an assistant watch our pipeline or summarize weekly debriefs without stealing anyone's focus?"

This chapter is a bit of an outlier in that it will explore what *could* be possible for you. You can go a long way in AI implementation without the use of agents and assistants. But I think you should know what's happening in the AI world and what you could accomplish, should you choose to take the next steps. In this chapter we'll explore:

- *Where agents already fit* with routine tracking, qualification upkeep, market intel, and feedback summaries.
- *What makes them safe* using boundaries, oversight, and human gates that keep control where it belongs.
- *How you can start small* by piloting a single, repetitive task and letting the results prove the case.

You might find some of these possibilities a little intimidating, making you feel like you need to catch up. But you're not behind. You're precisely on time to understand what's around the next corner.

And the good news? The same curiosity and caution you've practiced so far are exactly what will make you good at this next stage, too.

Where agents can help today

Most people still picture *AI assistants* as the stuff of tech demos or glossy keynote slides: chatty bots that promise to run your calendar, book your flights, and maybe overstep a little along the way. In reality, what's emerging inside professional firms is far more practical and far less dramatic. We're seeing dozens of small, specialized helpers that quietly handle the repetitive background work no one has time for. Think of them as *continuity keepers* instead of *decision makers.*

Here are a few places they already add meaningful value:

1. **Market intelligence that never sleeps**

In most firms, market research happens in bursts. We dive into the internet just before a proposal, or when a client lead surfaces. Agents flip that pattern by working constantly and consistently in the background. They can monitor open data sources, scan for published RFPs, or flag when a target client posts a new project or press release.

They don't predict the future; they simply *notice earlier.* That early awareness lets you act like a firm twice your size, because your intelligence loop runs all the time, not just on demand.

2. **Proposal tracking and follow-up**

A classic bottleneck in business development is version control as you try to keep track of which proposal went out when, who reviewed it, and what happens next. A lightweight assistant can log every submission, tag it by client and discipline, and notify you automatically when follow-up is due.

Instead of facing that "Did anyone hear back from the city?" moment, you'll receive a reminder in your dashboard before anyone has to ask.

3. **Qualifications upkeep and credentials management**

Keeping team resumes, project sheets, and past-performance data current is no one's favorite job, which means it's often neglected until the night before a deadline. Agents can watch for updates in your PM or HR systems and suggest corrections proactively.

"Three new projects were marked complete this month. Would you like to update your experience list?" Small nudges like that prevent last-

minute scrambles and keep your digital library proposal-ready year-round.

4. Feedback and debrief summaries

After presentations or client meetings, valuable insights often scatter across notebooks and inboxes. An assistant can gather notes, emails, and chat messages, then draft a short internal recap while the conversation is still fresh. Humans still edit and interpret, but the first 80 percent of the organizing is done.

The payoff from these small improvements isn't just efficiency, it's memory. The firm starts to preserve its activity and collective knowledge more reliably over time.

These aren't pie-in-the-sky scenarios. Every capability listed above already exists, inside software you may already use such as CRMs, scheduling platforms, or collaborative workspaces now extended by large-language-model plug-ins. The difference is intention, as you decide to connect a few of those features and let them run continuously.

And you don't need to hire a "chief AI officer" to get started. Start by watching where information currently leaks: the missed follow-ups, stale content, lessons learned then forgotten, and mistakes repeated. Those are the easiest places to train an assistant. Not because they're glamorous, but because they're tightly scoped and low-risk.

Use the same short "reality-check" meetings you already hold to keep tabs on how these tools are behaving. Make AI oversight a standing five-minute item—just enough to ask what's working, what's drifting, and whether any adjustment is needed. That light, recurring rhythm replaces the need for a single 'AI officer' and keeps responsibility distributed but visible.

The moment you stop losing time to repetition, you'll feel what these new systems are really about. It's not speed for its own sake, but *continuity*. Work that is remembered and integrated into the firm's collective knowledge.

The 4Rs of benefit

Once you've seen what these simple assistants can do, it helps to understand *why they matter*. Their real power isn't in any single task. Instead, they shine because they improve how your firm learns and operates every day. You can think of their benefits in 4Rs: *Routine, Response, Repetition, and Reflection.*

Routine: Reducing the friction of everyday work

Every firm has invisible procedures that burn time. These typically include weekly status reports, pipeline updates, reminders, and recaps. By shifting those to an assistant, you remove the mental drag of remembering the basics. Routine automation doesn't replace thinking. As with the automation of any menial task, it preserves energy for higher judgment. The firm starts each morning with small wins already handled.

Response: Delivering timely nudges

Deadlines rarely slip from neglect. They slip because no one spots the delay in time. An AI assistant, however, will notice when a file hasn't moved, or when feedback is overdue, and will act as a gentle tap on the shoulder. These micro-responses keep the momentum moving forward without expensive and annoying management overhead. The faster you can respond to small delays, the less you need to fix big ones later.

Repetition: Turning practice into consistency

A major benefit of AI agents is that they never tire of doing the same tasks the same way. They don't get bored in their work. That consistency prevents knowledge decay between people and projects. When a process is regularly repeated, such as logging proposals, formatting resumes, or summarizing meetings, it can be dehumanizing and demotivating for the team member who's responsible. But an AI assistant shines at repetitive tasks and builds standardization silently in the background. Over time, that reliability becomes part of your firm's brand.

Reflection: Capturing what you learned from the work

Every project produces lessons, but they often fade as soon as the next job begins. An assistant that gathers close-out notes or feedback summaries creates an instant archive of insight. People can then review, filter, and apply those lessons to the next pursuit. This is how small data becomes institutional wisdom.

Together, these 4Rs don't just increase efficiency, they build organizational memory.

- Routine keeps motion steady
- Response keeps teams alert
- Repetition keeps quality stable
- Reflection keeps people learning

When these elements converge, AI stops feeling like a quick utility and starts functioning as a real member of the workflow. It's quiet, reliable, and always ready. The result isn't more automation. It's better continuity, anchored by human oversight.

Governance and gatekeeping

Once assistants begin helping behind the scenes, a new kind of discipline becomes essential. Not technical discipline, but *managerial* discipline. The goal isn't to rein in creativity. It's to make sure the system's initiative always serves the intent of the firm and its clients.

Think of governance as good scaffolding. It supports growth without limiting it. Every firm that experiments with automation eventually learns the same lesson: when everything is automatic, you run the risk of nothing being accountable. The cure isn't fear, it's clarity.

1. Define who decides what

AI agents can notice patterns, propose actions, and even draft next steps. But they can't understand consequences. That boundary line should be explicit.

Write it down if you need to, and even make it an integral part of designing the agent in the first place: *"The agent recommends; a person approves."* Whether the task is sending a client summary, posting a report, or updating a database, decide in advance where the machine hands the baton back to a person.

2. Keep transparency visible

Every suggestion an assistant makes should leave a clear trail. You must be able to know when it acted, what data it used, and who confirmed the result. That small habit builds trust. People are far more comfortable when they can see what the system saw and how it reached its suggestion with no hidden processes and no invisible judgments. Just as you sign off on cost reports or drawings, you should be able to audit what your assistants are doing on your behalf.

3. Use version control as a safety net

Just as you do with your drawings and specifications, you must keep track of the evolution of any ideas. If an AI agent writes, edits, or compiles material, make sure the original, and all previous versions are well labeled and just a click away. Keeping "before" and "after" copies prevents the anxiety of loss and invites comparison.

4. Create light-touch checkpoints

The best oversight feels natural, not bureaucratic. A five-minute "assistant review" in staff meetings, for instance, can surface small surprises before they become big misunderstandings. *What did the system flag this week? Did any result seem off?* Little practices like that anchor the human presence and oversight in automated tasks.

5. Protect private and client information by default

Most contemporary agents rely on stored data for continuity. That continuity must never come at the cost of confidentiality. Keep any sensitive details, including client names, contracts, and financials, off-limits unless you know exactly how and where the data is being stored. As you would with any subcontractor, assume accountability for what the assistant touches.

Governance isn't red tape. It's peace of mind. When you make boundaries explicit, experimentation can expand safely. Assistants stay where they belong, remaining helpful, quiet, and fully auditable, while your people stay confident that nothing is happening without their awareness.

That's the balance that defines mature digital leadership: *agents that suggest, and humans who decide.*

***

Designing your first pilot assistant

By this stage, you probably don't need convincing that AI assistants can help. What you need is a safe, visible *first step* that helps everyone see the value without opening any risky doors. That's what a *pilot assistant* can do. It will give you a low-stakes trial that shows people what's possible, builds familiarity, and gathers evidence you can act on.

Start with a question, not a tool

Don't begin by asking, *Which platform should we buy?* Instead, ask, *Where do we consistently lose time or continuity?* The right pilot emerges when you trace small frustrations among those routine actions that no one loves but everyone needs. Think *proposal logs, meeting summaries, qualification updates,* or *weekly task check-ins.*

Look for tasks that share three traits:

1. *Repetitive:* They happen often, sometimes daily or weekly.
2. *Structured:* They follow a predictable format or set of steps.
3. *Containable:* They have clear boundaries and limited impact if they go wrong.

Those are your best test cases because success will be clear, and risk will stay minimal.

Frame the pilot like an experiment

Treat your first assistant test exactly as you would a design pilot or quality-assurance check:

1. *Objective:* What problem are we testing it against?
2. *Inputs:* What data or materials will it use?
3. *Boundaries:* What is it *not* allowed to do?
4. *Measurement:* How will we know if it helped? (Time saved? Task completed earlier? Fewer version conflicts?)

Write these four points on one page. Post them where everyone involved can see them.

Keep the team small and hands-on

Start with two or three volunteers. Ideally your team will include one "digital explorer" who's genuinely excited, one "steady skeptic" who keeps you honest, and one project manager who can translate outcomes into business terms. That mix keeps enthusiasm grounded in reality and ensures both excitement and caution are represented in the feedback.

Record and review out loud

Each time the assistant runs, capture what happened. What did it do well, what was misunderstood, what do you wish it could do next? Five minutes of note-taking per cycle is enough. After a week or two, discuss those insights in a standing meeting or quick lunch huddle. Hearing the trial described aloud turns invisible progress into shared learning.

Declare a finish line

Set an end date for your pilot that is, at most, two to four weeks out. By then, you'll know enough to answer three simple questions:

1. Did it save measurable time or effort?
2. Did it create any new risks or confusion?
3. Would it be worth refining or expanding?

If the answers are yes, no, and yes, congratulations, you've found an early success worth documenting. If not, that's fine too. Your firm just earned a low-cost lesson about where automation *doesn't* add value yet.

Celebrate the evidence

When you see a win, make it visible. Post it on your intranet, share it at the next staff check-in, or add it to your "Quick Wins" whiteboard from Chapter 12. A well-designed pilot converts curiosity into conviction. It gives skeptics something to inspect and supporters something to build on.

A successful first pilot doesn't prove you're "AI-enabled." It proves you're *AI-ready.* Your firm is now able to test, measure, and learn with purpose. That's how small experiments evolve into sustainable systems.

A short note on the "how"

Just to be clear, this book and this chapter isn't a guide to building or programming AI agents, and it's not meant to be. My aim is simply to show you what's possible so you can make informed decisions about where to explore next. It's not to turn you into a developer overnight.

But, having said that, you don't have to be a developer to set an assistant in motion. Most modern systems already include an automation layer. You just need to connect it to a clear, repeatable task.

Here are three accessible paths that you could begin with:

1. **Native tools you already use**
   Many common platforms now have built-in assistants or "co-pilot" functions: *Microsoft 365 Copilot, Google's Duet AI, Adobe Firefly*, or *Notion AI*. These can watch shared folders, summarize documents, or remind you to update content on a schedule. Starting here means using tools your IT team already trusts.

2. **No-code workflow builders**
   Services like *Zapier, Make (Integromat),* or *Notion Automations* let you connect different apps. For instance:
   "When a new file is added to the "Proposals" folder → send it to an AI summarizer → post the summary to our internal chat channel."
   That's an AI-assisted workflow, in other words, an early agent. And it can be created without writing a single line of code.

3. **Specialized "agent" platforms (for later)**
   As your comfort level grows, you may be interested in testing tools such as *ChatGPT Teams, OpenAI Assistants API*, or modular frameworks like *Anthropic Workflows* or *Microsoft Power Automate with GPT*. These can maintain context across sessions and execute recurring tasks. You don't have to tackle this stage until your simpler pilots are running smoothly.

If this all feels a bit too technical, remember, you don't have to create and wire the system yourself. You just need to identify *what task* deserves automation, then *pair up* with your firm's most curious tech lead or consultant who can translate your workflow into a small, testable loop.

Start simpler than you think. Even a once-a-week reminder generated by AI to refresh project data qualifies as a pilot assistant. You'll learn far more from one reliable, visible success than from a half-finished grand design.

Ethical connections

When the technology starts to run quietly in the background, it's easy to forget that its actions still carry your professional signature. Every document sent, every update logged, every reminder issued bears your firm's (and therefore, your) name. That's why hands-off automation still demands *hands-on accountability.*

Here are some guidelines that can help keep you safely between the guardrails.

1. Ownership is never optional

No matter how autonomous a system feels, responsibility doesn't shift to the software. If an assistant sends a draft, updates client data, or builds a report, ultimate accountability rests with a human being who verifies, approves, or supervises the output. Write that expectation into policy, and repeat it out loud during project reviews. Automation can lighten the work, but it cannot absorb the duty of care that comes with it.

2. Be transparent about where AI is used

Clients, partners, and even colleagues deserve to know when a piece of work included help from an AI system. Transparency builds trust, while secrecy erodes it. You don't need to publish technical details, but you can acknowledge that certain background tasks (summaries, reminders, or document organization) are automated. The message you send is simple: *we use technology to improve quality and consistency, not to avoid responsibility.*

3. Stay vigilant about data boundaries

Efficiency should never tempt you into cutting corners on consent or confidentiality. Limit each assistant's access to only the data it needs, and be explicit about what's off-limits such as client contracts, financials, and proprietary designs. Keep those boundaries visible and reviewed regularly by whoever owns your firm's data governance. Just because software can connect systems doesn't mean it should.

4. Keep people in the conversation

An ethical AI culture depends on inclusion, which means inviting everyone, not just the tech enthusiasts, to shape how automation fits into daily practice. Ask how it changes workloads, affects opportunities, or alters staff expectations. Listen especially to those who worry about displacement. Often, they're pointing to areas where communication or training need to catch up.

5. Remember that ethics scales with impact

The moment an agent or workflow touches client-facing material, the risk to your reputation multiplies. That's why ethical reflection isn't a one-time setup; it's a standing agenda item. As systems grow more capable, revisit your commitments to privacy, attribution, labor transparency, and accuracy. A quarterly "AI check-up" works as well as any audit.

Why this matters

Your firm's reputation rests not just on what you deliver, but on *how* you deliver it. Sure, AI can handle routine work invisibly, but your integrity still shines through in how decisions are made, credits are shared, and boundaries are respected.

The firms that succeed with AI won't simply automate faster. They'll also be recognized for their *responsible adoption*. They'll be using technology to amplify professional judgment rather than replace it.

In the context of AI and AI agents and assistants, ethical considerations should not be thought of as a brake to stop the progress of your integration. Instead, you should think of it as the steering wheel.

Try this now: The two-week agent pilot plan

By now, you've learned what agents are, where they can help, how to manage them, and the ethics that keep them safe. Your next step will be hands-on. Not theoretical, not aspirational, but something you can literally start next Monday.

This short "Try It Now" section gives you a framework for testing a small assistant in just two weeks, using what you already have. This exercise isn't about the tool. It's about your *practicing* with it.

Week 1: Plan and prototype

Day 1 – Choose the task

Pick one small, repetitive process you already control. Choose something that takes time but not creativity. Examples might include:

- Compiling weekly meeting notes
- Logging proposal submissions
- Sending follow-up reminders after client calls

If it happens often and follows clear steps, it's a good candidate.

Day 2 – Define success

Write one sentence that completes this line:

"The assistant will save time by __, and we'll know it worked when ___."

Clarity prevents you from getting off track and just one measurable outcome is enough.

Day 3 – Map the inputs and outputs

List what goes *in* (files, emails, messages) and what should come *out* (a summary, alert, or update). This will make setup simple when you build the workflow.

Day 4 – Pair with a platform

Decide *how* to run the task:

- Try an assistant already embedded in your firm's software (like *Microsoft Copilot* or *Notion AI*).
- Or use a no-code connector such as *Zapier* or *Make* to automate the steps.

The goal is function, not perfection.

Day 5 – Pilot run #1

Test once, watch what happens, and jot down what went right and what confused the system. The value of Week 1 is *learning the edges.*

Week 2: Refine and review

Day 6-9 – Adjust and observe

Run your assistant three or four times this week. The minor adjustments that will be quickly obvious, like phrasing a summary prompt more clearly or filtering input folders, will teach you faster than big overhauls. Keep notes: Where does it help? Where does it stumble? Does it cause new friction elsewhere?

Day 10 – Share the results

Hold a short "demo debrief" with anyone touched by the process. Show before-and-after examples, summarize the time saved (or wasted), and capture comments (both positive and skeptical). This conversation ensures the pilot stays *with* the team, not just with the tech-curious few.

Your two-week debrief template

You can use this simple recap to close out the experiment:

What we tested	What worked	What didn't	What we'll try next	Owner
Lead-tracking reminder	Saved 2 hrs/week	Missed 1 deadline alert	Refine trigger rules	Jenna

Small pilot projects will accumulate in value faster than you might expect. After just a few well-documented pilots, you will begin to see a clear pattern of progress.

Maintaining your momentum

At the end of the two weeks, either refine the assistant, retire it, or replace it with a new candidate. The discipline is the same each time: *choose, test, measure, learn.* If you keep that rhythm alive, you'll gradually shift from *experiments* to *ecosystem* as you evolve to a workplace where digital helpers are simply part of how good work gets done.

Launching your first agent doesn't mark the end of your learning, it marks the start of a new kind of partnership. From here on, the skill isn't about mastering every new tool. It's about stewardship in which you guide the technology that remembers, supports, and grows with you.

CHAPTER 14

Ethics and Trust: What clients expect

The trust gap in AI adoption, or, Why disclosure equals credibility

When firms first begin integrating AI into their client-facing work, the biggest risk isn't technical error, it's *perception.* Clients aren't just buying the quality of your proposals or deliverables. They're buying confidence in the people that stand behind them.

AI unsettles that exchange because clients rarely know what "using AI" actually means. A firm says it "leverages AI," and a client wonders:

- *Is my confidential information being uploaded somewhere?*
- *Who is really doing the work, a human or a machine?*
- *What happens if the AI makes a mistake? Who's accountable?*

These questions reveal the *trust gap,* which is a growing distance between what firms *intend* when they adopt AI and what clients *perceive* when they hear about it.

Why the gap matters

In relationships built on reputation and judgment, which is the essence of the architecture, engineering, consulting, legal, and financial professions, clients equate *transparency* with *reliability.* They don't expect you to avoid new technology, but they do expect you to *own how you use it.* Silence, on the other hand, sounds like secrecy.

The firms gaining the most ground today are those that lead with disclosure rather than dodge it. A simple statement such as: *"We use AI tools to help assemble data and format quickly, but every analysis and recommendation is reviewed by our team before it reaches you,"* does more for your credibility than any case study or capability slide. By satisfying their curiosity, you prevent their anxiety.

What full disclosure actually communicates

Openly stating when and how AI is part of your process communicates three crucial ideas:

1. *Competence.* You know how to handle new technology responsibly. Disclosure shows that you're in control, and not merely experimenting (playing) with a new toy.
2. *Integrity.* You have nothing to hide, which allows clients to see you as a partner who values clarity over convenience.
3. *Confidence.* You know that your human judgment still drives the outcome. By framing your use of AI as an assistant, not a replacement, you reinforce trust in your people.

How to bridge the trust gap

Start small, but make it visible. Add a single sentence of disclosure wherever you already talk about process. This might be around your conversations, website content, or proposal statements about quality control, risk management, or data security notes. Keep it in plain language and avoid legalistic jargon. For example: *"Our team uses AI tools for routine document preparation and analysis under strict confidentiality and human review."*

That single line reframes the conversation. Instead of fearing automation, clients understand your leadership in supervising it.

AI won't cost you a client because you're using it. It will cost you if they *discover* it later and feel that you have concealed it. In professional practice, disclosure isn't a confession, it's your next competitive credential.

What to say when you use AI

Your conversation with clients about AI doesn't have to be technical. In fact, it *shouldn't* be. The moment you sound like you're defending a system instead of explaining a process, trust begins to wobble. What clients really want to hear is that your work still has a person in charge, and that your tools make that human even better.

Lead with clarity, not disclaimers

You don't have to open every proposal or meeting with an "AI disclosure statement." Instead, you can simply weave acknowledgement into the way you describe your team's approach.

- Instead of: *"This report was prepared using advanced AI systems to deliver faster analysis."*
- Try: *"We use AI tools to accelerate research and improve formatting, but every conclusion and recommendation is produced and verified by our team."*

The second phrasing sounds like partnership, not replacement. It assures your audience that technology is a means, not a mask.

Translate "AI-speak" into plain language

Your clients aren't asking for a tutorial, they're asking for reassurance. Replace jargon or your own internal shorthand with words that connect directly to value and responsibility.

Instead of saying…	Say it this way…
"We leverage generative models to produce documentation."	"We use modern AI tools to prepare first drafts more efficiently."
"Our workflow integrates automated summarization."	"We use automation to help organize materials so our team can focus on the analysis."
"This model improves throughput on data streams."	"It helps us handle data faster without compromising accuracy."

Think about the *function*, not the *feature*. When a client can picture what happens and who's in control, they stop worrying about the black box.

Use "we" statements

Where AI can feel impersonal, people retain the warmth. Maintain the person-to-person connection by keeping the human subject in every sentence. This means writing in the first or second person and staying away from the passive voice by using phrases such as:

- *We use…*
- *Our team reviews…*
- *Each recommendation is considered by…*

Even subtle phrasing such as "we use" instead of "it generates," reminds clients that their trusted relationship remains with people.

Match your message to tone and context

- *In proposals:* Keep it to just one line, integrated with your quality or methodology language.
  "We use AI tools to speed document preparation and ensure consistency; every deliverable is reviewed and refined by our project team."
- *In meetings or interviews:* Keep it relaxed and conversational.
  "We've found that AI helps us organize information faster so we can focus on deeper analysis and creative problem-solving."
- *In ongoing relationships:* Reinforce it subtly during projects.
  "An AI assistant tracked version updates for this report. It saved us a few hours and reduced errors, so our review time could be spent focusing on the recommendations."

Repetition and regular inclusion of AI in all your conversations normalizes the practice. Over time, it shifts AI from a curiosity to a sign of competence.

The tone test

Before using any AI-related description, run this quick test: *"Would I say this comfortably to our most cautious client—the one who values familiarity and trust over novelty?"* If the answer feels hesitant or defensive, simplify the wording until it conveys confidence and calm control.

The language you use when discussing you firm's use of AI-assistance isn't a script to be memorized. It's an extension of your firm's voice. You're not explaining a *tool*, you're demonstrating your *judgment.* Clients don't expect perfection, they expect your presence and they want to work with people who remain accountable even as the work becomes more digital.

Privacy and data boundaries

Trust deepens when your clients know that their information is treated with care equal to your own. For most professionals, confidentiality has always been second nature and the trust of your clients has been a foundational element in your practice. But AI introduces new places where information can travel, replicate, or be stored outside your immediate control. Being transparent about how you protect data is therefore both a technical safeguard *and* a marketing advantage.

Clients hear "AI" and imagine leaks

Even if your firm's systems are secure, clients may picture their project details being fed into a global database, or a chatbot learning from their private documents. Those mental images aren't paranoia—they're the result of high-profile tech stories. The smart response isn't to tell them not to worry, it's to show, simply and visibly, how protection actually works.

Four everyday commitments to privacy

1. **Limit what the system can see**
   - Never upload an entire proposal, contract, or drawing set into any platform without reviewing where that data is stored.
   - Use redacted samples or structured inputs, such as specific excerpts instead of full archives. This isn't just security hygiene; it shows clients that you practice minimum exposure by design.

2. **Keep client identifiers out of prompts**
   - When working with generative tools, remove names, locations, or proprietary details. Say *"Client A: municipal park redesign"* instead of *"City of Maplewood Waterfront Park."* Important habits like that enforce confidentiality while letting assistants support the work safely.

3. **Work only within approved environments**
   - Use enterprise or professional versions of AI tools with agreed-upon data policies and opt-outs from model training. If your IT or legal team hasn't vetted a product, treat it as off-limits for live client data.
   - You can even note this commitment in marketing materials: *"Our firm uses only enterprise-grade AI systems where client data remains private and is never shared outside secured environments."*

4. **Keep a human checkpoint before anything leaves your network**
   - No AI output—draft, analysis, or summary—should ever go directly to a client without having been reviewed by a person. That quick inspection not only protects quality but also ensures no sensitive metadata or hidden remnants slip through.

Explain data care the same way you explain design care

You don't need to scare or lecture. A few calm sentences in a proposal, pitch, or onboarding conversation signal both your competence and your integrity:

"We apply the same confidentiality standards to AI-assisted tasks as we do to every other part of our workflow. Client data stays within our secure environment, and no material is shared or stored on public systems."

This kind of phrasing replaces mystery with method. It reassures clients that your digital growth hasn't changed your professional discipline.

The no-surprises rule

Ultimately, privacy practices come down to one principle: *never let technology introduce surprises into trusted relationships.* Every time you explain how data is handled (before being asked, of course) you shorten the distance between innovation and trust. Security by itself can make you compliant. But security combined with free and open client communication builds confidence.

Internal accountability

Accountability doesn't end when a file leaves your firm's server. While AI tools can assist, they can also make errors quietly and quickly. Imagine an outdated figure being inserted, a citation omitted, or a paragraph phrased in a tone that your brand would never use.

Professional trust isn't defined by whether mistakes happen, but how quickly and clearly you own them. When work involves AI assistance, that ownership needs to be explicit, shared, and visible, especially inside the team.

Clarity begins at the task level

Every team using AI should be able to answer three questions before a project starts:

1. **Who reviews the output?**
   In the same way that you must have an individual sign and seal your drawings, there must always be a named reviewer for your AI-assisted output. We're not talking about a department, but a real person who confirms that any AI-assisted result meets your firm's quality and ethical standards.

2. **Who signs off?**
   Review and approval are not the same thing. Approval signals final accountability, even when review is distributed. When everyone knows who signs, the process earns respect.

3. **Who records the lessons?**
   Each AI interaction offers a small learning opportunity. Maintaining a short "AI log," in which you document what worked, what failed, and how corrections were made, strengthens institutional memory and guards against errors being repeated.

Correcting the machine is part of leading it

In the old workflow, one person corrected another person's draft work. In the new workflow, a person must also correct the machine's draft. That doesn't dilute professionalism; it redefines it.

Treat every correction as evidence of human judgment at work. If an AI summary misstates a timeline or merges two projects incorrectly, note both the fix and its source. Over time, this discipline creates a shared internal library of what your systems tend to misunderstand. That's how oversight evolves into training.

And always, either feed those corrections back into the same conversation or system that produced the draft, or add them to the global context prompt that you use to orient AI in future conversations. Most AI tools "learn" locally within a session or workspace thread rather than universally. If you start a new thread, the system forgets the previous feedback. Maintaining continuity, either by referencing prior guidance or keeping a running project context, helps the AI retain relevance and reduces repeated errors.

Make accountability visible, not punitive

Accountability isn't about assigning blame. It's about keeping competence traceable. Simple measures, such as initials beside reviewed content, metadata tags identifying the last person to edit the content, or a short note in project files such as *"verified for accuracy by ___,"* keep authorship transparent without slowing the team down.

Clients rarely see these signatures, but they feel the result. The work is cleaner, corrections are faster, and confidence is clearer.

Leadership's role in shared responsibility

Leaders model accountability every time they ask: *"Who reviewed this?"* rather than *"Who created this?"* That shift signals that quality is a collective value, not an individual risk. It also builds psychological safety for the team. People become more comfortable working with AI assistants when they know human oversight is both expected and respected.

From ownership to assurance

The fact that the people on your team own the outcomes doesn't mean that you distrust the technology. It means that you know that your professional ethics are maintained regardless of who or what has created the content. When every draft, decision, and correction still passes through a named professional, clients can relax, knowing that your accountability is not diluted by digital help.

The message this sends, internally and externally, is simple: *AI may assist your process, but people remain responsible for the promise.*

Ethics as brand insurance

Ethics rarely appears on a balance sheet, yet it protects more value than any insurance policy. In the age of AI, choosing where *not* to automate, disclose, or rush, has become one of the clearest signals of credibility. Clients can feel the difference between a firm that experiments responsibly and one that chases efficiency, novelty, or the latest toy at any cost.

Reputation is the return on integrity

Reward comes to the firms that stay steady while others over-promise. When your proposal language, project updates, or public statements consistently show discretion and care, clients internalize one simple message: *you are trustworthy*. That impression compounds. A single mis-step with client data or an over-automated proposal can undo years of confidence, but steady transparency builds a dividend of reliability that keeps paying off.

It's the same principle behind the "goodwill" that is calculated and valued during a firm's sale. It's the intangible value of that firm's reputation and relationships. In practice, ethical restraint is one of the ways that goodwill is built.

Modern ethics as risk management

For professional services, brand risk doesn't begin with what AI *does*, but with how responsibly people *appear* to use it. Every internal decision, from prompt sharing to assistant setup, contributes to shaping that perception. Before releasing any AI-assisted material, ask three short questions:

1. *Will the client immediately understand that people remained in control?*
2. *Could this workflow expose or misrepresent any private information?*
3. *Would I stand behind the result if it landed on tomorrow's front page?*

If the answer to any of those feels uncertain, your proper ethical response is restraint. True professionalism often looks like patience.

Restraint as a differentiator

Adopting AI into your workflow doesn't demand that you push every boundary at once. In fact, clear boundaries become competitive assets. When a client observes that your firm turns new tools on *selectively*, and only in areas that you've tested and proven to be safe and useful, they read that as maturity, not hesitation. They see a team confident enough to balance innovation with judgment.

As more firms automate similar tasks, what will set yours apart isn't the tool you use, but how visibly you govern it. Ethics becomes your brand's insurance because it protects one vital trait that AI cannot reproduce: trust built over time.

Making ethics visible without preaching

You don't need slogans or policies taped to the wall. Visible ethics is simply *behavior clients can see*:

- Including a short note about data boundaries in your proposal boilerplate.
- Referencing human review in footer notes on your deliverables.
- Discussing lessons from an AI-assisted task at a client meeting.

Each of those actions whispers, *"We think before we automate."* Over time, that whisper becomes your firm's reputation.

The payoff

When a client hires you, they're betting their own reputation on yours. Your consistent record of ethical transparency gives them—and you—safe cover in volatile conditions. It's why the most advanced firms are often the most cautious. They innovate publicly, but only after they've tested privately and documented deliberately. The restraint they use isn't hesitation; it's confidence they've built from the control they've exercised.

In a noisy digital marketplace, the loudest differentiator may soon be calm reliability. And the firm that practices ethical restraint today will not just own the trust others spend years trying to rebuild, they'll avoid the kind of misstep that can erase that trust in a heartbeat.

***

Try this now: Create an AI use disclosure checklist

Ethics becomes real only when it's portable and every member of the team can carry it into a meeting, a draft, or a conversation. A simple internal checklist turns the abstract idea of "responsible use" into repeatable behavior.

Why a checklist matters

Airline pilots with decades of experience still rely on a checklist for every flight. They know that it isn't bureaucracy. It's reliable memory. A checklist keeps your judgment consistent, especially when projects move fast, or staff turns over. The purpose isn't to restrict creativity. It's to make sure every client sees the same discipline, no matter who prepares the work.

Five questions to anchor your firm's AI checklist

1. *Disclosure* – Have we clearly stated, in plain language, where and how AI has supported this deliverable? If not, add a short, human-sounding note before it leaves the door.
2. *Confidentiality* – Have we confirmed that no proprietary or client data was entered into non-approved systems, or left exposed in public tools? If uncertain, recheck before submission. One review is cheaper and easier than one explanation.
3. *Human review* – Has a named person reviewed, corrected, and approved the AI-assisted content? If not, pause. Accountability still requires a signature.
4. *Tone and discretion* – Does this output sound like *us*? Remove any machine-leaning phrasing or over-confidence that could misrepresent judgment.
5. *Recordkeeping* – Have we logged this instance of AI use, including any lessons learned or prompts worth saving? What you track today becomes the training foundation for tomorrow.

How to make it live

- *Keep it visible.* Post the checklist or its condensed version near proposal templates or project-launch forms.
- *Make it conversational.* Encourage small teams to treat it as a five-minute talk at the end of deliverable review: *"Did we check these items?"*
- *Revise it quarterly.* As tools and data policies change, update the language. The act of updating reinforces awareness more than the document itself.

Using the checklist with clients

You don't have to hand clients the list, but you *can* use it to guide transparent language. You might, for instance, add a short sidebar or footnote in proposals that reads: *"Our firm reviews all AI-assisted content through a defined quality and confidentiality checklist before release."*

A line like that tells clients your ethics are operational, not aspirational.

What this final step accomplishes

By installing a disclosure checklist, you convert the three themes we've talked about in this chapter, transparency, protection, and accountability, into one daily habit. It closes the trust gap not with policies, but with predictable behavior. Over time, this practice becomes part of your firm's identity. It becomes the invisible standard that turns digital competence into reputational strength.

From disclosure to judgment

While the transparency we've been talking about will build trust, it's your ongoing judgment about how you use and assimilate AI that will sustain it. Once your clients believe in *how* you work, their continued confidence depends on *how well* you decide. The next challenge for design professionals isn't whether to use

AI, but how quickly you can apply discernment when the tools accelerate every task.

This chapter established your habit of ethical visibility in which you consistently show clients what you're thinking. Our next chapter shifts focus to what happens within that thinking. With digital speed, we're easily tempted to skip the pause that makes insight possible. We're going to learn how to keep your professional judgment clear, calm, and accountable.

CHAPTER 15

Teaching Judgment : When not to automate

Why judgment still rules

There's a justified pride that comes with experience. It comes from that sense that after years of projects, clients, and deadlines, you can *feel* when something's off long before you can explain it. That feeling has always been the intuitive instrument of leadership and professionalism. And in an era flooded with dashboards and data, it's also the instrument most at risk of being drowned out.

AI has brought pattern recognition to near-superhuman levels. It can see correlations in seconds that once took analysts weeks. It sees which words improved a win rate, which subconsultants speed delivery, which proposal sequences tend to fail. To the uninitiated, it can look like pure foresight. Patterns appear, and decisions follow.

But what machines perform as *pattern*, humans perform as *judgment*. And those two are not the same species of intelligence.

Patterns aren't promises

AI recognizes what *has* happened and then it predicts what *usually* happens next. Judgment asks, *"Should it happen again?"*

The distinction sounds philosophical but it's immensely practical. Where pattern recognition turns probability into guidance, human intuition turns context into wisdom. A pattern might reveal that hospitals with faster elevators tend to earn higher satisfaction scores. Judgment asks whether investing in elevators belongs at the top of the next budget cycle, or whether that pattern simply masks deeper design issues like wayfinding or daylight.

Without that interpretive leap, pattern is at risk of becoming prophecy and data can begin to sound like destiny. In business and in design, both of these are a poor substitute for informed and intelligent choice. Pattern and data can assure you that, *"This worked last time."* Judgment pauses to ask, *"I wonder if that's still true."*

Intuition as pattern memory

Some executives like to describe intuition as "gut feeling," as though it lived somewhere below the diaphragm. In truth, intuition is memory operating faster than consciousness. It's the brain's internal pattern library that has been built from triumphs, scars, all-nighters, and awkward client calls that taught lessons no spreadsheet could measure.

That intuition isn't mystical. It's encoded experience, distilled through reflection. The difference is that your internal model was trained not merely on data, but on consequence. It's the lag between cause and effect. It's the client's body language after a risky comment, the contractor's correction three weeks later, and the email that diffused tension. Those moments taught you the nuance that machines still can't feel: *what it means when humans react.*

Every professional with the slightest bit of emotional intelligence has spent a career noting and labeling those subtleties, quietly building a private dataset that no algorithm can access. It was the weight in a client's voice, the tone of a pause, the look in a reviewer's eyes. These signals are invisible to any sensor, but they've been decisive in every project and pursuit.

While AI may sense the patterns of words and data, you sense the mood in the room and the intangible yet very real 'vibe.' That's judgment.

The risk of pattern intoxication

Once you see how accurate the models can be, the temptation to trust the dashboard more than the discussion becomes very real. Teams begin to look for validation inside graphs instead of conversation. What was once instinctive risk assessment becomes spreadsheet compliance.

That shift feels scientific. But it can be deceptively dangerous because business development, marketing, and design are not closed systems. They live amid psychology, politics, and timing, the hot mess of human variables that no pattern ever fully contains.

There's an old saying from aviation: *"The plane flies right up until the moment it doesn't."* Autopilot works wonderfully until the universe throws a new variable into the system. A bird strike or a sudden windshear. An error the sensors never saw coming. And that's when the pilot's experience and calm judgment matters, precisely because they know when to seize back the controls.

In your firm, that "bird strike" might be a client crisis, a public backlash, or a team misunderstanding that's more emotional than factual. AI won't see it coming, because it's not a data anomaly, it's a human one.

Judgment is your equivalent of manual control. It's what keeps trust, ethics, and empathy flying when the screen flashes an answer that doesn't feel right.

The enduring advantage

The firms that thrive in this next era won't be those that obey every algorithm. They'll be the ones that remember which decisions must pass through a human pulse. Machines can manage the information, but only people can decide what truly matters. Computers can read a resume and predict probability of success. Only a human leader can read the moment and sense *what's at risk for this person, right now.*

That subtle comparison—pattern recognition versus situational intuition—is your real professional advantage. Anyone can buy the same analytics subscription. You're the only one who can apply your awareness of when a subtle, silent pause means doubt, respect, or that something has gone wrong. Judgment is not the enemy of data; it's the conscience of it. It doesn't resist automation; it supervises it.

As you introduce AI deeper into your workflows, keep one guiding principle in mind: *Let the data inform you, but let your humanity decide.* Because long after dashboards predict our next move, clients will still hire the person who knows when not to trust the dashboard, and when to trust their own intuitive certainty instead.

The 3 checkpoint model: Purpose, Emotion, Reputation

Judgment doesn't come from genius. It comes from taking the moment to pause. Between the speed of data and the impulse to decide lies a moment of stillness that professionals once took for granted. That pause is what kept the work grounded in purpose, not panic. Now that decisions move at machine tempo, that pause has to be made visible. It must be taught, named, and protected.

Here's a simple way to build it back into the daily work of your firm: Teach your team that every decision must pass through three checkpoints; *purpose, emotion, and reputation.* Think of these as three short internal questions to be asked before any automated process gets to run on trust.

Checkpoint 1 – Purpose: Why are we doing this at all?

When AI or automation offers a suggestion, the first responsibility of leadership is to reconnect it to *intent.* You pause and ask: *What human purpose is this decision meant to serve?* That question reframes every algorithmic *"because we can"* into *"should we?"*

Automation can easily take activity that isn't really strategic and make it seem irresistible. A pursuit assistant might generate twenty near-identical outreach emails, achieving tremendous volume but totally lacking in meaning. A data model might optimize for click-throughs instead of comprehension. By checking for purpose, you bring the focus back to, *"Does this make the work more valuable to someone we care about?"*

Before approving any new automated step, name the purpose out loud. If the team can't say who benefits and how, the task stays manual.

Purpose clarifies alignment. The moment the purpose feels thin or self-referential ("because it's faster," "because the software can"), that's your alert that human thought needs to retake control.

Checkpoint 2 – Emotion: What will this feel like?

Every decision has an emotional outcome, whether the spreadsheet measures it or not. So, the second checkpoint asks: *"How will this choice land with the people who experience it?"*

In this case, emotion isn't sentimentality; it's risk recognition. Projects, proposals, and communications travel through the very human nervous systems of clients, staff, and reviewers. Each of them comes with their own point of view, laced with anxieties and hopes. When automation strips tone or nuance, it runs the risk of saving time at the cost of trust.

Conduct a simple test by reading the draft, email, or output aloud. Listen for the emotional temperature. Does the piece reassure or escalate? Does it sound informed or indifferent? These quick empathy scans catch errors that no grammar checker ever could. If a mechanized or automated tone makes a person feel unheard or disposable, the efficiency of the tool has overshot the wisdom of the people.

Emotion is the middle gate because it reminds you that even correct data can wound if it's delivered without care. While machines can measure engagement, only people can sense relief.

Checkpoint 3 – Reputation: Who will we be after this?

Every output, automated or not, leaves a small fingerprint on your firm's standing in the community and among your peers and clients. The final checkpoint asks: *"If this decision were shared publicly tomorrow, would we still be proud to have our name on it?"*

Your reputation is viewed in the rear-view mirror. It's built on what you've done in the past. But protecting that good standing requires that your judgment look forward. Today's shortcuts can't be allowed to undermine tomorrow's credibility. Most automation errors that make news aren't technical. They're ethical blind spots that are revealed when the public light shines on them. The damage is always measured in lost confidence.

Leaders can practice this checkpoint by narrating it aloud in meetings: *"Before we automate that, let's think about how it reads to a client seeing it for the first time."* That question reframes optimization into stewardship.

When your team learns to ask that kind of question instinctively, you've taught true digital maturity. You've moved from efficiency for its own sake to efficiency aligned with integrity.

Putting the model to work

In practice, these three checkpoints don't slow decisions; they right-size them. Add them to one-minute pauses in your workflow:

1. *Purpose:* Is this solving a real problem for a real person?
2. *Emotion:* How will it feel when they receive it?
3. *Reputation:* Will this reflect who we want to be known as?

Every time you walk a team through those gates, you turn judgment from a private talent into a teachable process. Over time, that process becomes integrated into your culture. It becomes part of the collective muscle memory that keeps human sense guiding machine speed.

Judgment, like strength and trust, increases through repetition. The more often you ask these questions, the less often you'll need to repair the consequences of skipping them.

Common failure points

Even the best systems should include a warning label that says: *"Faster doesn't always mean wiser."*

As AI becomes more embedded in everyday work, its greatest strengths—speed, scale, and accuracy—can quickly become its blind spots. Not because the tools misbehave, but because busy professionals begin trusting them a little too easily. The challenge isn't mechanical; it's psychological. Automation can alter our sense of pacing by making *activity* feel like *progress.*

Speed over sense

Anyone who's experimented with AI knows the temptation. You ask a model for a draft, and it delivers something polished in seconds. The cursor blinks, the syntax is clean, and before you've even taken a sip of coffee, you've got a finished document. In that moment, the urge to press *send* can be overwhelming.

The momentum and sense of accomplishment can feel intoxicating. But unverified acceleration can be racing you straight toward a cliff. When speed outruns scrutiny, subtle errors can easily sneak through. Mismatches in tone and false assumptions can sneak in with implications that a client will notice faster than you will. You haven't gained speed. You've traded professionalism for motion.

Speed becomes dangerous when it disguises *avoidance.* The first draft looks so good and so polished, who wants to go back and verify data and sources? That's where judgment must reach for the brake because slower doesn't have to mean stagnant. Sometimes the most productive moment in a project is the pause between the machine's reply and your next keystroke when you ask, *"Does this actually move us forward?"*

Healthy tempo in the digital era doesn't mean perpetual forward motion. It means deliberate rhythm: adapt, review, release. Teams that master that rhythm still use AI's acceleration, but they finish their day with the certainty that it's been done right, not merely fast.

Data over dialogue

Another hidden hazard is mistaking reporting volume for understanding. Dashboards, metrics, and predictive analytics have become the new vocabulary of progress. They track engagement by the decimal and visualize trends with amazing precision. Yet every chart, however elegant, is still an abstraction of the human conversation that created it. When a team starts quoting numbers but stops asking questions, data has replaced dialogue.

Dialogue is where nuance lives. It's in the clarifying follow-up from a client, the side comment in an interview that reveals the real priority, and the tone of an email that analytics will never chart. Those fragments of conversation are messy but priceless. They supply the missing nuance that raw data can't generate.

In healthy firms, numbers initiate discussion; they don't end it. If an assistant reports that proposal engagement is down 20 percent, the question shouldn't be *"What can we automate next?"* but *"Who can we talk with to understand why?"* Because almost every quantitative shift hides a qualitative story. AI can summarize that story, but only people can *listen to it.*

When leaders begin to be frustrated that listening takes too long, the culture hardens and curiosity fades. Soon, meetings and decisions revolve around metrics, and the clients—the very humans behind those metrics—fall down the list of priorities. Emphasizing dialogue isn't old fashioned, it's essential for good judgment.

Restoring balance

The correction to both illusions—speed over sense and data over dialogue—is remarkably similar: *reinstate conversation before confirmation.*

Before approving, sending, or automating, invite a question or a colleague by asking, *"Does this sound right to you?"* or *"What am I missing here?"* That micro-moment of dialogue re-anchors your professionalism. It re-humanizes the process exactly where the risk of drifting off course begins.

Where machines work in straight lines, people work in messy, overlapping circles of understanding. Those circles, including their questions, clarifications, and reflections, are what keep technology from turning your professional competence into dangerous complacency.

Building judgment into team culture with *Story, Debrief, and Reflection*

Judgment doesn't scale because someone issued a memo. It spreads when people watch how others, especially those in authority, are behaving. People learn discernment the same way they learn design or leadership, not by memorizing rules, but by watching someone else handle a tough moment well. It's caught before it's taught.

When firms talk about "embedding AI responsibly," they often reach for new policies, workflows, or training. Those are important, but culture rarely changes through documentation alone. Culture evolves through *stories, shared reflection, and thoughtful debriefs* that turn experience into learning fuel. That's how personal judgment becomes integrated into organizational intelligence.

Start with stories

Every team already has its legends. The pursuit that barely came together but taught everyone to handle pressure differently. The client who surprised you with a last-minute curveball. And how, despite the unreasonable demands, the project manager kept her composure. These are not just anecdotes. They're the actual neural wiring of your firm's culture.

The goal is to retell those stories explicitly through the new lens of human-plus-machine work.

- *"What judgment did we show here that no algorithm ever could?"*
- *"Where did speed tempt us to skip understanding? And how did we recover?"*

By labeling good calls and near-misses as *examples of judgment*, you give teams a vocabulary for reflection without having to preach it. The story is fully capable of conveying the lesson.

1. **Debrief, don't autopsy**

After major pursuits or project milestones, schedule short debriefs that sound like curiosity sessions and are *not* blame sessions. Avoid asking *"What went wrong?"* and substitute *"What did we learn that we want to remember?"*

Three questions can turn any debrief into an engine for sharper judgment:

a) *What signals did we notice early that proved important later?*

b) *Where did automation help—or risk misleading us?*

c) *What would we watch for next time that isn't in the data?*

Keep them short, recurring, and conversational, just fifteen minutes, standing up, if possible. That rhythm communicates that reflection is a normal part of the work, not extra work. Over time, these short reflections turn good judgment into second nature. They remind everyone that AI tools are partners, not pilots. They also teach younger staff that *asking reflective questions is a strength, not a rookie hesitation.*

2. **Reflection as an everyday skill**

Reflection doesn't have to wait for the big pursuit wrap-up. It can be baked into daily moments. A pursuit manager can end a quick huddle with: *"Before we break, what judgment did we show today that a machine couldn't?"* A marketer reviewing AI-generated copy can ask aloud: *"What did this get right, and what does it still not understand about our message?"*

These one-minute pauses train discernment faster than any formal workshop. They signal that thinking and questioning is an integral part of output quality.

3. Leaders model the pause

Good leaders teach good judgment less through their speeches and more through what they do under pressure. A principal who says, *"That result looks great, but let's sit with it for a minute before we send it,"* is modeling courage in their restraint. That sentence tells the team, *"We value accuracy over ease, integrity over instant response."*

Even more, when leaders identify their pause: (*"Here's what I'm double-checking and why,"*) they give clear permission for others to pause, too. Over time, that visible composure becomes contagious, Calm and reflection will spread faster by demonstration than policy.

Workshops: When a machine should pause for a person

The more you and your team practice judgment in real situations, the more dependable it becomes when you need it. It's one thing to value discernment in theory. It's another to feel it, live, when a real decision clock is ticking, and an AI assistant is insisting that it knows best.

That's why teams should practice judgment in context with short, low-risk scenario workshops that make people think: *"When should we let the model run, and when should we step back in?"* These sessions don't require fancy software or new budgets. They require leaders who know that reflection can be a training tool, not a delay.

The purpose of a scenario workshop

The goal isn't to expose mistakes. It's to sharpen instinct before the stakes are high. Think of it as a flight simulator for professional ethics and common sense.

A good scenario workshop teaches three reflexes:

1. Recognizing when automation is out of its depth.
2. Identifying what kind of human information is missing.

3. Deciding who should re-enter the loop, and how fast.

If participants come away with better questions, not just better answers, the session succeeded.

Designing a simple workshop

Start with everyday material, something that is familiar enough that no one feels defensive. Choose a *realistic but not critical* scenario that AI could plausibly touch:

- An AI summarizer drafts an email that inadvertently over-promises scope.
- A predictive ranking tool suggests skipping a long-time subconsultant because of low "efficiency" scores.
- A proposal generator uses language that's technically correct but emotionally tone-deaf.

Step 1: Read the output

Let the group look at what the machine produced. No commentary yet; just reactions.

Step 2: Spot the flags

Ask participants to call out what feels *off*. Encourage both intuition and rationale. You'll hear comments like:

- *"The tone's too casual."*
- *"That sentence assumes something we'd never promise."*
- *"It misses what this client values."*

Those flags are your teaching moments. They reveal where human cues still matter.

Step 3: Ask the judgment questions

Use the 3 Checkpoint Model we explored earlier as your framework:

Purpose:	Why does this step exist? Who benefits?
Emotion:	How will it feel when received?
Reputation:	If this went public, would we be proud of it?

Listen for insight beyond the surface fix. The richest discussion often begins where someone admits, *"I'm not sure. Let's talk it through."*

Step 4: Rewrite or re-decide

Have the group improve the output using both human judgment and AI assistance. Ask, *"How can we teach the model a little of what we just learned?"* This reinforces AI as partner, not scapegoat.

Step 5: Reflect aloud

End the session with a one-sentence takeaway from each participant:

- *"Next time, I'll slow down before assuming the first draft is right."*
- *"Now I see how small word choices affect trust."*
- *"I need to check what data the model is using before believing its confidence."*

These quick reflections lock in understanding more powerfully than lectures ever could.

Making it a habit

Hold these workshops monthly or after any major pursuit. Keep them lighthearted but serious in purpose. Over time, they will normalize the language of judgment just as safety briefings normalized risk awareness on job sites. Soon, you'll hear the cultural shift mid-conversation:

"Let's pause the automation until we've run the checkpoints."

"That was a machine decision. Who's the human sponsor on this one?"

That's the sound of a firm where discernment has become integrated into operations and culture.

The real takeaway

Critical thinking doesn't slow your progress. It safeguards the trust your clients have invested in you. A firm that practices judgment under calm conditions will respond decisively when pressure comes. When the unexpected hits with a client crisis, a misstep on social media, or an AI model that's been hallucinating, people who've rehearsed the pause won't panic. They'll know instinctively when a machine should pause for a person.

Try this now: The weekly "Pause Moment" reflection

Good judgment doesn't stay sharp on its own. It becomes dull when you go too fast. Which is why even the most well-intentioned teams need a rhythm for slowing down, if only for a few minutes, to notice what the week's flurry of automation has taught them. The "Pause Moment" is a simple way to keep that discipline alive. It takes ten minutes a week. No slides. No metrics. Just a handful of real conversations about what felt *right* and what felt *off* when AI entered the workflow.

How the Moment works

1. **Schedule a Friday pause**

   Pick a consistent time, maybe Friday morning before deadlines or Friday afternoon before the weekend. The predictability matters more than the duration. It signals that reflection isn't a luxury. Like QA/QC it's smart maintenance and smart practice.

2. **Ask three simple questions**

   Go around the table or team channel and invite brief answers to:

   a. *Where did automation save us time this week—and did the result still feel like us?*
   b. *Where did we override an AI suggestion, and what did that tell us?*
   c. *What's one judgment call we want to remember for next time?*

3. **Keep it conversational.**

   Make sure everyone knows they're not expected to deliver a polished presentation. A few plainspoken seconds of truth are far more useful than a lengthy report.

4. **Capture one line of learning**

   An ideal takeaway might be nothing more than:

- *"We learned that faster summaries still need human tone."*
- *"We noticed the model favored old data. Next time we'll check recency."*

These brief records build a living library of experience. This will build into your firm's collective intuition about using AI, written in plain words.

5. **End on reputation, not speed**
   Close this little practice by asking, *"Did our choices this week reflect who we want to be known as?"* That one question raises the reflection from workflow efficiency, where AI shines, to cultural identity, where human judgment can never be replaced.

Why it matters

The point of these "Pause Moments" isn't more meetings for their own sake. It's to calibrate your use of this remarkable new tool. It's how you'll keep your compass aligned before the pace of production bends it. These moments re-center purpose and empathy before errors need correcting.

Over time, the Pause Moment will build confidence across your entire team. Where the obvious and predictable conversation around AI is about speed, your people will begin to use the word "pause" in conversation, not as a slowdown but as an act of responsible leadership. They'll have learned that responsibility sometimes means waiting an extra minute before trusting the screen.

The lasting advantage

If every firm has access to the same data (which AI ensures they do) the only durable difference is *how thoughtfully people choose to use it.* That's the competitive edge your good judgment protects.

So, end each week by taking that breath, however brief. Run the checkpoints. Reflect aloud. Listen for the tone in the room. Because the future of trust won't belong to the fastest firms. It will belong to the ones whose people know when to pause, look up from the screen, and decide like people again.

PART 5
SUSTAINING THE ADVANTAGE

CHAPTER 16

Learning loops: Debriefs for continuous improvement

Why learning loops matter

Every firm starts strong with new tools because early progress feels exciting. The first AI experiments make people curious, hopeful, and sometimes even a little proud that they're keeping up. But once the novelty wears off, improvement becomes less about discovery and more about discipline. That's where most firms start to lose ground.

The difference between quick adopters and lasting innovators usually comes down to one thing: *the habit of learning while the work is still fresh.*

Your team already knows how to debrief a pursuit or project, but those discussions often stop at the surface. We note what went wrong, what worked, and who stayed late. Learning loops go deeper. They close the circuit between *what we did*, and *what we saw*, by adding *what we'll do differently next time.* They make sure that the reflection doesn't just fade with the next deadline.

In human terms, a learning loop is simply a structured pause that asks, *"What did this teach us and how will we use that lesson tomorrow?"* When you build that question into your routine, you start building a reliable advantage. You spot problems early enough to fix them, and wins early enough to repeat them. You notice patterns in feedback, in client reactions, and in the way your tools behave.

AI tools make this even more critical. Because when automation accelerates the work, it also accelerates the spread of small errors. A single misguided prompt can ripple through dozens of documents before anyone notices. Regular learning loops are how you spot those ripples early and turn them into improvement instead of rework.

This practice isn't complicated. It means finishing a pursuit or project with the same curiosity you began it with. It turns reflection from something we do *after* the work into something that *sustains* the work. Over time, those brief conversations teach both the team and the systems that support them. Human judgment becomes the right kind of feedback loop. It's one that is rooted in awareness, not autopilot.

The anatomy of a loop

A learning loop is simply a short, reliable process for turning experience into progress. Instead of being a meeting, a report, or another dashboard, it's a habit. It's a way to make sure each round of work teaches the next one something useful.

Every complete loop includes four simple stages:

1. Feedback
2. Insight
3. Adjustment
4. Re-test

Together, they build a rhythm of learning that fits naturally into real work instead of sitting beside it.

1. **Feedback: What actually happened?**

Every loop begins with noticing that some things went well, some things were missed, and some things just felt off. The first step is to capture that feedback while it's still fresh. It could be a client comment, a teammate's observation, or a pattern the data revealed. Keep it short. A single sentence is enough:

- *"The summary hit all the points but sounded mechanical."*
- *"We missed the client's tone in the follow-up email."*

In this stage, you're not judging; you're recording. The goal is to preserve the raw material before memory fades or narrative takes over.

2. Insight: What does it mean?

Next comes reflection. This is the step most teams skip. This is where we ask, *"Why did this happen?"* and *"What is the lesson here?"* This is where the difference between correction and learning becomes clear. While correction will fix the task, learning improves your approach going forward. Correction might suggest, *"Let's rewrite that paragraph."* But learning takes it a step beyond by saying, *"Next time, let's check the consistency of tone before sending."*

Insight connects the instance to a principle. It helps the team understand their behavior, not just the end product.

3. Adjustment: What will we do differently now?

This is where you turn the insight into a concrete change. That could mean updating a checklist, tweaking a prompt, adding a review step, or shifting ownership of a task. The adjustment should be as small and specific as possible. Big reforms rarely stick; small ones usually do.

Note that, beyond simply correcting an error, you're teaching your workflow to avoid the same error next time. And your AI-assisted workflow is learning, one move at a time.

If the insight was about an AI output, this is also where you feed the lesson back into the system by editing a custom instruction, saving a better example, or documenting a good prompt in the team library. You're not punishing the machine; you're teaching it.

4. Re-test: Did the change work?

A loop isn't closed until you try the new approach and see the effect. That might happen naturally on the next project, or intentionally in a quick practice task. The point is to check, *"Did our adjustment make the result better?"*

If it did, great. Capture that as a reusable pattern. If it didn't, you've just refined your understanding, which is still progress. Either way, the loop ends with learning, not guessing.

Making the loop routine

Learning loops only work when they're small enough to fit inside normal days. You don't need a committee or extra agenda item, just the discipline to finish the week by asking those four questions:

1. What happened?
2. What did it mean?
3. What will we change?
4. Did it work?

This doesn't add work, it finishes it. Over time, teams who run these small loops find that improvement stops being a special project and becomes part of how the firm thinks.

From debrief to dataset

Good teams hold debriefs. Great teams make those debriefs searchable. The difference isn't the quality of the conversation; it's what happens afterward.

A learning loop ends best when what was learned doesn't disappear into memory. Turning a short discussion into a small, structured note makes it useful later, especially when patterns start to repeat. This doesn't mean you need to build a database or introduce more forms. It simply means capture what you learn in plain language, in one consistent place.

Start simple

At the end of each debrief or learning loop, write down three things:

1. *Situation:* What it was about.
2. *Observation:* What happened or changed.
3. *Adjustment:* What we'll do differently next time.

That's enough structure for the human brain to recall the pattern later and for the AI tools in your workflow to detect trends if you decide to collect them. For example:

1. *Situation:* Proposal for hospital renovation.
2. *Observation:* The summary read as too technical for a non-clinical audience.
3. *Adjustment:* Simplify language and add two empathy checks before delivery.

That's only three sentences, but they'll be priceless when another healthcare pursuit begins.

Keep it where people work

The best reflection notes live close to the tools your team already uses, not on a forgotten drive. If your people work in *Teams*, *Slack*, or *Basecamp*, make a *"Learning Loops"* thread or channel and post short recaps there. If you use project management software, tag debrief notes to the project. The goal is convenience: it has to be easier to store a note than to forget it.

Look for patterns, not perfection

As the library grows, spend a few minutes each quarter scanning through it. The point isn't to grade teams; it's to see where the same notes keep showing up. If several projects mention "tone too formal," you've found a teachable theme. If different groups describe the same automation error, you've found an update the tech lead can address. When you track learning this way, improvement stops being random. It becomes recognizable.

Add a little help from AI

Once you have a modest log of lessons, simple AI tools can help sort and surface them. Ask an assistant to group entries by topic or detect recurring words. You're not analyzing with heavy data science. You're using the same everyday language to highlight trends. For instance, the model might tell you: *"In the last six months, three notes mention unclear roles in QA review."*

That single observation can inform training or workflow tweaks faster than any formal audit.

Make the loop visible

When people see their input appear later, either quoted during a training session or influencing a new guideline, they understand the loop is working. That kind of recognition closes the circle between *sharing* and *improving*.

If lessons that have been filed away are never referenced again, both reflection and enthusiasm for the process fade quickly. But when leaders bring a few of them back into meetings: *"Last quarter we noticed this pattern, so we're adjusting it here,"* you reinforce the idea that reflection changes things.

Regular debriefs become a collective memory system as one project's correction turns into another team's advantage. That's how small notes become a knowledge asset. Rather than being stored for formality or because we believe we're supposed to, they're revisited and reused for momentum.

Human context first

Once teams begin to formalize their learning loops, there's always a temptation to make them neater. Someone suggests tracking themes in a spreadsheet or assigning scores to improvements. The idea comes from good intentions. After all, people want to see and measure progress. But it can pull the process away from what matters most. Learning loops succeed because they're human. They capture the feel of a situation, the voices in the room, and the small cues that pointed to a better choice. When those moments turn into metrics, the meaning that made them useful starts to disappear.

A note such as *"the tone felt off"* or *"the client seemed uncertain"* might look vague on paper, yet it carries important context. It's a clue about trust, timing, or emotion; none of which a machine can score. That context is what helps the next team interpret the lesson, not just repeat the correction. So, whenever a loop produces new learning, keep a sentence or two that explains the human side of what happened. For example: "*We simplified the summary because the client's feedback sounded hesitant, not critical. They needed reassurance more than detail.*" That simple explanation ties the adjustment back to empathy, which is the real skill every firm needs to preserve.

Human context also comes from naming the conditions around events. These might include the pace of the week, who was under pressure, or what expectations shaped the decision. A pattern of small misses may tell one story in a busy season and a different one when workloads are light. The loop works best when it remembers that people's state of mind is part of the data set.

It's worth repeating in team meetings that learning loops aren't performance reviews. They're memory builders. When people know their observations will be treated with curiosity, not as evidence, they're far more likely to share them honestly. Leaders can reinforce that tone by going first. The boss is sharing something they noticed in their own work and how they adjusted. That kind of openness keeps the focus on growth rather than fault.

When reflection keeps its human details—how it felt, what it looked like, and what surprised us—it stays real. The result isn't a neater dataset; it's a stronger sense of understanding. That's what allows lessons to transfer from one project to the next without losing their texture.

In the end, human context is what keeps a learning loop from collapsing into a checklist. Numbers will confirm that something changed, but the stories explain why it mattered.

Codifying the win

The kind of reflection we've been talking about matters most when it changes what happens next. Every insight, whether it came from a client comment, a debrief note, or an AI misstep, has value only if the team can reuse it without having to rediscover it later. Codifying a win simply means taking what worked and making it repeatable.

Turn moments into guides

Most improvements begin as something someone says aloud:

- *"That prompt worked better when we added examples."*
- *"Shorter subject lines got more replies."*

Those small discoveries lose power when they stay locked in the conversation where they first appeared. Capture them in plain language somewhere easy to find: inside an internal playbook, a shared chat channel, or a quick-reference document. Note what changed, why it helped, and when to use it again.

This isn't about adding process. It's about saving thinking time. The next person shouldn't have to guess what the last person already solved.

Teach the system what you've learned

When an AI tool is part of the workflow, a "win" can also become a simple configuration update. Save good examples in a prompt library, revise custom instructions with what you've learned, or label strong outputs so the assistant has better reference material. Every time you feed the system a small example of good judgment, you extend your team's expertise into the next project.

Over the course of writing this book, as I've collaborated with my AI assistant, I've logged hundreds of tiny corrections that I've both kept notes about and "conversed" about with the bot, so that the adjustments have become part of its ongoing instructions. These have included comments about focus, sentence structure, syntax, point of view, and dozens of other small, but important aspects of writing the manuscript.

In your case, that collection of refinements built up over months, becomes your firm's unique operating layer. It's gone far beyond a simple prompt to become a blend of process memory, tone guidance, and preferred phrasing that reflects how you actually work. It's how the machine begins to sound like your culture instead of a generic data model.

Share the lesson, not the template

Codification works best when it feels like storytelling, not policing. Whenever a team captures a useful improvement, share it briefly in the same way you'd pass along good advice.

- *Here's something that helped us handle a tough reply.*
- *Try this phrasing next time—it landed well with the client.*

When people see these notes circulate casually, they start contributing their own. A culture of sharing forms naturally when wins feel like collective property rather than compliance.

Avoid the trap of over-systemizing

It's possible, and even tempting, to make something *too* formal. If every new insight requires filling out a form or waiting for system approval, the reflection slows down. Keep the process light with short notes, short updates, and visible results. The goal is to maintain momentum, not to build a database. A conversation followed by one short sentence in the playbook is enough.

Keep it alive

Review your accumulated "wins" now and then to confirm they still represent good practice. As conditions or technology change, some guidance will expire. Editing a living library shows that your learning is continuing to evolve.

When teams see their own refinements built into the firm's tools and templates, they understand that their daily learning directly shapes the enterprise. Lessons stop being personal anecdotes and become collective intelligence.

Codified wins are the bridge between one-time reflection and operational consistency. They prevent teams from having to learn the same lesson twice and make each project a tiny rehearsal for the next. Over time, that growing library of smart adjustments becomes one of your firm's most durable assets.

Introducing "Skills" and assistant routines

At some point, your growing library of lessons will start repeating itself. Patterns emerge in the notes, in team conversations, even in the questions people ask. When you see the same judgment being applied again and again, that's your cue that the learning isn't just personal anymore. It's ready to become a skill the whole firm can use.

What a "Skill" really is

In this book, "Skill" doesn't mean learning a new software or taking a refresher course. It's shorthand for a small, teachable pattern you can hand off to a person, a checklist, or an AI assistant. A Skill is a bit of embedded experience. Examples might include:

- A polished tone framework that keeps proposal language consistent
- A short review checklist that catches emotional temperature before delivery
- A short prompt sequence that helps an assistant summarize client notes with empathy

Each Skill encodes one trustworthy piece of judgment that a team discovered or previously lived only in someone's head.

Capture patterns as micro-routines

Every time the team says, *"Remember to do it this way,"* that's a potential Skill. Write it down and name it. Keep the format short: *Purpose, Steps, Caution.*

- *Purpose:* what this helps avoid or achieve.
- *Steps:* two or three quick actions or phrases that anchor the result.

- *Caution:* the condition that reminds people when review by a person is still required.

Use ordinary language. A Skill titled *"Tone Tuning for Client Emails"* might read:

- *Purpose:* Keep follow-up emails sounding personal, not procedural.
- *Steps:* Start with gratitude, restate one related detail from the meeting, then confirm next action.
- *Caution:* Always re-read before sending.

That's all the structure you need. Keep the list simple enough for anyone to use without explanation.

Teach it to the assistant, too

Once a pattern proves reliable, give it to your digital helper the same way you would brief a new team member. Save the phrasing, preferred wording, or checklist as part of the assistant's custom instructions. Point it to one or two model outputs that capture the tone you want. Little by little, your tools will begin to consistently reflect the team's collective sense of judgment.

If you've already been logging micro-corrections or enhanced prompts, you're halfway there. Each refinement you captured earlier can now live as a defined Skill:

- *Check emotional tone on client communications.*
- *Format summaries with a human voice, not marketing copy.*
- *Highlight assumptions in data before presenting recommendations.*

These short, teachable anchors will continue to evolve as your firm's use of AI does.

Keep people in the loop

Even as your assistants learn, they need a steady human hand. Think of digital Skills the way aviation thinks of autopilot: It's very helpful for cruising at altitude, but always under supervision. During reviews, ask simple validation questions: *Is this skill still helping? Is it keeping our tone aligned?* This way you'll know when to update or retire one.

In best practice, you'll refresh your Skill sets quarterly, pruning what's outdated and adding fresh judgments from new projects. This keeps the human-plus-machine partnership balanced.

Why it matters

When you convert steady reflection into shared Skills, you're building a culture that keeps improving on its own. Each project contributes something small. It could be a sentence, a tone cue, a Skill, or a better workflow. But together those pieces create an institutional reflex for doing things right.

The goal isn't full automation. It's continuity. A new hire can learn from yesterday's smart adjustment. An AI assistant can echo the firm's preferred voice. Done this way, future projects will always begin one step wiser than the last one ended.

That's what a mature learning loop looks like: people and tools refining each other until improvement becomes part of how the organization thinks, not just something it talks about.

Try this now: Your first learning loop session

You don't need to wait for a major project to start. The small, low-stakes moments are where learning loops take root. The goal of this exercise isn't to create a new meeting, it's to build a reflex.

Step 1: Pick a recent moment (not a crisis)

Choose something normal from the past week. It could be a proposal draft, a client conversation, or a short task that used AI. Skip the high drama and start with an everyday example where things mostly worked but could be sharper.

Step 2: Spend ten minutes together

Gather the two or three people who were involved. You can do this standing at someone's desk or in a quick video chat. Set a timer if it helps. Promise each other to talk honestly but briefly. Then ask four questions, the same four that form the core of every loop:

1. *What happened?*
2. *What did it mean?*
3. *What will we change?*
4. *Did it work?*

Keep the tone conversational. You're not writing a report, you're staying curious.

Step 3: Capture a note before you move on

Record one or two short sentences before logging off:

- *We learned that tone checklists help new writers keep focused on the client.*
- *Adjust AI summary prompts to include one sentence of empathy.*

Post the note where the team already lives, such as your shared channel, PM tool, or a Learning Loop thread. The act of writing it down is what closes the loop.

Step 4: Name a potential Skill

During the conversation, listen for anything that sounds like: *"We should always…"* or *"Next time, remember to…"* That's the outline of a Skill in the making. Give it a temporary name and keep it on a short list. You'll refine it later once it's been tested again.

Step 5: Reflect on tone and momentum

End with a light self-check:

- *Did this feel helpful or perfunctory?*
- *Was it brief enough to make repeating it easy?*
- *Did anyone feel unheard?*

These answers are the first test of the loop's design. If it felt natural, you're on the right track. If it dragged, shorten it. Reflection survives only when it fits comfortably into real work.

Step 6: Schedule the next one

Rather than declaring a "program launch," just say, *"Let's do this again next week."* Regularity is what turns the behavior into culture. Before long, you won't need an invitation. The loop will happen by instinct, in hallway chats or quick project reviews.

The strength of a learning loop lies in its simplicity. Run one small session this week, capture one insight, and carry one change forward. That's enough to start a self-improving rhythm, which has always been the hallmark of firms that keep evolving long after the novelty of a new tool fades.

When your internal learning loops begin working smoothly, the benefits soon reach beyond the firm. Clients start to feel it in the consistency of your tone, the steadiness of your follow-through, and the reliable confidence that every project builds on the last. Continuous improvement isn't only an operational habit; it's a form of care. It tells clients that you're still listening even after delivery day, still learning from their reactions, and willing to adapt with them.

That mindset becomes the foundation for our next chapter—*Smart Client Care*—where reflection turns outward, and your firm's curiosity about its own work becomes empathy for the people it serves.

CHAPTER 17
Smart Client Care

The shift from CRM to Care

You can feel it in nearly every conversation with clients lately. The tools keep getting smarter, but the relationships somehow feel thinner. Firms can now log every touchpoint, automate every thank you, and track engagement down to the minute. And yet, when clients talk about the partners they trust most, the word they use isn't "efficient." It's "thoughtful."

That gap between *management* and *care* is the real frontier. Over the past twenty years, Client Relationship Management systems (CRMs) were heralded as the cure for inconsistency. They promised that if you just captured the data, you'd preserve the bond. In truth, most systems (to the degree they've been used at all) ended up measuring contact instead of connection. They told us *who* we reached and *when*, but not *how it felt* when we did.

Now, with AI embedded across nearly every client-facing platform, that risk has been multiplied. The same automation that can draft a perfect follow-up note in three seconds can also drain it of personality in three milliseconds. The line between attentiveness and autopilot has never been thinner.

So, we're going to begin this chapter with a reset. *Smart client care* is not the next evolution of CRM; it's the correction to it.

Remembering, not reminders

Clients don't want to be managed. They want to be remembered. They want to see signs that your firm noticed the small details. For example, the comment they made in passing, the hesitation behind their feedback, the pressure they were under in a public meeting last month.

A phone call that says "I saw the press release" means very little if the next sentence doesn't reflect awareness of what that announcement actually meant to them.

In that sense, the most advanced relationship system you can build is still the human nervous system. Digital tools should exist to *extend* its reach, not replace it.

Smart client care is like the barista who starts your order before you ask. It's not about tracking; it's about noticing. The system holds the memory, but the feeling of being recognized still comes from a person.

When you think of it that way, all the automated touchpoints, including emails, quarterly check-ins, and project updates, stop being entries in a database and become one more chance to say, "*We still know you.*"

Continuous Relationship Management

If we keep the letters but change the meaning, CRM could just as well stand for *Continuous Relationship Management.* Continuous doesn't mean constant messaging; it means *unbroken context.* The information collected during a pursuit ought to inform the relationship long after the ribbon-cutting. Notes from a project debrief should feed the next kickoff meeting, so the client never has to repeat what they've already taught you.

Imagine a client lifecycle that feels seamless rather than segmented:

- A marketing coordinator captures a tone preference ("Keep language conversational").
- An AI assistant surfaces that note months later when drafting a proposal follow-up.
- A project manager sees the same cue before presenting design options.

No one person had to remember; the system remembered *for* the people, but always in the service of empathy rather than efficiency.

That's what separates smart care from smart software. It's the psychological continuity that tells a client, *"You didn't start over with us. You picked up where we left off."*

The human signal inside the system

A CRM full of names means nothing without emotional metadata. By that I mean the tone, urgency, and mood behind each entry. AI, when guided with care, can start to detect those signals, grouping updates by sentiment or surfacing moments of silence that might need a call rather than a campaign. But human meaning has to train the model. A note such as, *"Client sounded tense about the schedule"* may look casual, but it's the clue that prevents a misunderstanding later.

That kind of notation feels unscientific, yet it's the soul of professionalism. Smart client care uses digital precision to preserve emotional memory. The system simply holds what the people observed, so empathy doesn't have to suffer from the drag of turnover or forgetfulness.

In a world obsessed with scaling relationships, this is how real relationships scale. It's not by increasing frequency, but by amplifying attentiveness.

The reframed role of technology

Used wisely, the new generation of CRM and AI tools should make empathy easier, not rarer. Automated summaries can help you recall what was said, analytics can remind you when to follow up, and predictive alerts can flag when a client's priorities shift. But none of these functions replace the moments that turn a transaction into a bond. It's the handwritten note, the quick check-in, the gracious reply delivered before they had to ask that has the client characterizing you as "thoughtful," rather than merely "efficient."

Technology's job is to hold space for humanity, to catch what we might miss and to give us more time to be ourselves with the people who pay attention to that difference. A well-wired firm doesn't sound like a machine. It sounds like real people who are unhurried by one.

The next question follows naturally: if AI can keep the context, can it also help sense the emotion inside it? The answer is yes—when we teach it how. In the next section, we'll look at how intelligence becomes truly "smart" only when it learns from empathy, turning digital insight into human understanding.

***

Intelligence with empathy

While data may tell you *what* changed, empathy explains *why*. The two have to travel together, or the intelligence you gather about clients can quickly turn cold.

AI can process thousands of signals in an hour including emails, responses, open rates, and project updates. And it can deduce patterns that a person could never track alone. But none of that matters until someone interprets what the pattern means emotionally. A dip in engagement could signal disinterest, or it could mean your client's team was sent into crisis by a budget freeze. The algorithm can notice the silence, but only people can understand the story inside it.

That's why the future of client care isn't machine-driven insight, it's *humanized insight.* It's the kind that uses technology the way a flashlight uses batteries. The batteries provide power, but not direction.

Context before correlation

Smart care begins with context. When you ask an AI assistant to summarize client interactions, the output will always be accurate yet incomplete. It can tell you *how often* an account manager reached out, or *how quickly* the client replied. What it can't tell you, unless you teach it, is *how those moments felt.*

This is where small changes in habit matter. Instead of logging only factual updates *("Meeting held," "Proposal sent")*, add short emotional markers:

- *"Client seemed relieved by cost-savings idea."*
- *"Public meeting clearly weighed on the project director."*
- *"Excited about community impact. Let's use that phrase again."*

Those fragments of tone become the breadcrumbs that allow intelligence to mimic understanding. Over time, simple text-analysis tools can notice which kinds of emotional notes precede strong outcomes. They surface not just correlations but cues, which serve as reminders of the moods, stresses, and values that shape decisions.

AI isn't forming empathy itself because it can't. But it can help you map empathy so your team can act on it together.

Adding feel to the field data

In other industries, marketing teams often track something called *voice-of-customer* feedback from the thousands of comments, emails, and ratings that collectively reveal how people feel about a brand. To make sense of all that text, computers run what's known as *sentiment analysis*: a scan that tags each phrase as positive, negative, or neutral so the team can see mood trends at a glance.

That's an early, primitive version of empathy at scale in which machines are classifying tone by keyword. But professional-services work lives in a different emotional climate. A client's message is rarely pure praise or pure frustration; it's usually both in the same sentence. *"We're thrilled with the design—please don't let costs escalate again."*

A literal dataset would tag that as "mixed." A smart person reads it as *"We trust you; don't make us look foolish."*

Modern AI tools can now be trained to recognize this kind of nuance *if* you feed them examples labeled in human language. Each time your team annotates a client exchange with meaning: *"optimistic but cautious," "friendly tone, urgent timing,"* you are effectively teaching the system what empathy sounds like. Over time, dashboards stop reporting emotion as color codes and start echoing the insights your best relationship builders already sense intuitively.

From dashboards to dialogue

The real shift isn't technical, it's conversational. Instead of long reports no one reads, AI can condense months of correspondence into one simple statement: *"Across your top three clients, positive tone aligns strongly with proactive schedule updates."* That's a strong business case for increased empathy. It links behavior to outcome in a language that leaders and principals respect.

Smart firms then take that finding and ask the next human question: *Why?*

Maybe proactive communication makes clients feel in control. Maybe it signals reliability at a glance. Either way, the insight returns to where it always belonged, inside a conversation about trust, not metrics.

This is the feedback loop intelligent systems were built to support. Humans set the tone, AI mirrors it back, people refine it again. Together, the people and the AI turn intuition into something teachable.

Empathy as the new efficiency

Empathy used to be a side note in client management. It was an optional quality you hoped your best seller-doers possessed naturally. In a data-dense world, empathy has become available in every corner of your firm. It's a key operational tool that helps you prevent misfires before they cost projects, credibility, or even clients.

When your client-care intelligence is guided by empathy, efficiency improves without anyone feeling processed.

- AI drafts the reminder,
- You add the reassurance,
- The client feels both remembered and respected.

That's both efficient *and* human.

Ironically, as automation increases, empathy becomes measurable again. But it's not measured as a mood, but as a result in faster responses, smaller misunderstandings, fewer escalations. Empathy pays in reduced rework and steadier trust curves.

Real intelligence isn't about removing feeling from the client equation. It's about proving that feeling—that *touchy-feely* stuff that has made so many technical professionals cringe—makes business sense. And it's available to everyone, regardless of personality type, because AI has become your window on the client's emotional temperature.

Bringing empathy into the workflow

You don't need a new platform to begin bringing empathy into the client equation. You need a new question. Try adding a single line to every internal summary your team sends: *"How did the client seem?"*

It takes ten seconds, and it teaches the minds of the people, and eventually the machine, to connect information with emotion. After a few months of answering that question, your data library will show something extraordinary: *patterns of tone tied to patterns of outcome.* That's when your CRM stops acting as a ledger and starts acting as a mirror. Your AI will learn what matters because you'll keep reminding it that people matter.

Looking ahead

The next step in smart client care is sensing patterns in silence. There are always subtle warnings and quiet cues that predict dissatisfaction or opportunity before words have even been spoken. That's where *signals and sentiment* come in. They can recognize behavioral patterns not as data points, but as early messages from the relationship itself.

Signals and sentiment

Sometimes the most important message from a client is the one you never receive.

Every professional knows the moment: the inbox goes quiet, calls take longer to return, meeting notes feel perfunctory, or a phrase such as *"Let's circle back later in the year"* starts appearing more often. These are not transactions; they're *signals.* They're small shifts in timing, tone, or volume that reveal sentiment long before dissatisfaction becomes explicit.

AI can help surface these cues faster, but it takes the awareness of people to understand what they mean and how to respond.

Reading between the interactions

Traditional relationship tracking measures activity: how often you contact a client, how promptly they reply, and how satisfied they claim to be on a survey (if they completed it, and if they completed it honestly). But relationships rarely fracture right after a bad score. They erode through *quiet drift.*

That's where "signal-based" awareness comes in. By scanning the everyday traces of engagement, such as email cadence, meeting frequency, response time, and the tone of comments, modern tools can flag subtle deviation from a client's usual rhythm. For example:

- *A project champion who normally answers by day's end now takes two.*
- *Reading rates on shared documents drop by a third.*
- *Meetings that once ran an hour are now booked for half the time—a signal that attention is declining.*

Individually, these would mean nothing. Together, they suggest gravity has shifted.

When AI spots these patterns early, it doesn't hand you a conclusion, it hands you a question: *"Has something in their world changed, or is it something in ours?"* Either way, it matters.

Client sentiment isn't a score. It's a story.

It's tempting to translate those signals into numbers: *"72 percent positive," "moderate risk," "client health score 8.4."* Those dashboards look impressively scientific, but behind the decimals the empathy has disappeared.

In professional design services, *sentiment*, is narrative data. It's the running story of how clients feel about progress, partnership, and pressure. Like any story, it has plot twists including exhilaration in the early design phases, frustration during contract documents and negotiation, and relief at handover.

AI can identify those points if you feed it the right source material such as meeting transcripts, feedback text, and comment threads. But the interpretation must always remain with the people. When an algorithm claims *"sentiment downturn detected,"* a good relationship manager doesn't panic. They pick up the phone to find out if the context has changed. As long as the question stays human, the data remains useful instead of corrosive.

Acting on the small signals

Patterns only help if they lead to conversation. Here's how smart firms turn digital whispers into meaningful contact:

1. *Acknowledge the data privately, act publicly.*
   Use alerts as early warnings for your team instead of client-visible panic buttons. If responsiveness dips, gently re-engage through curiosity rather than correction.
2. *Correct through care.*
   Don't forward the "client health score" to staff with red exclamation marks. Instead, ask in your next internal huddle: *"What might be happening on their side that we should understand before our next touch point?"*

3. *Pair each quantitative clue with qualitative context.*
   When AI says engagement is down, have someone recall recent events in which you interacted with the client: budget planning, leadership change, election cycle, weather crisis. Empathy can fill the gaps that metrics leave behind.
4. *Respond visibly but naturally.*
   Sometimes all it takes is an unexpected note of appreciation. For example, a quick *"Thinking of you as the school project breaks ground this week"* is a small authentic gesture that can recalibrate tone better than a dozen status reports.

Over time these small acts turn signals into trust currency. The client learns that their subtle shifts are noticed and respected, not exploited. That feeling is what we really mean by being "responsive."

The emotional weather report

Think of client sentiment as today's weather, but not it's climate. Weather changes every day, but the patterns reveal the seasons. A healthy feedback loop measures both.

Using simple AI pattern recognition, firms can now build dashboards that look less like spreadsheets and more like *emotional weather maps.* They visualize intensity, movement, and trend rather than score. A storm in one region of the relationship can signal the need for shelter elsewhere.

Just as pilots read turbulence reports before flight, project leads can read sentiment maps before high-stakes meetings. They don't replace instinct, they sharpen it. When you walk into a conversation already aware that tone is breezier—or frostier—than usual, you begin with empathy instead of surprise.

Listening for silence

Of all the signals, silence is one of the most important that AI can quantify but humans must interpret. A model can detect longer gaps between communication or fewer data points from a familiar voice, but only you can sense *why*.

Some silences mean satisfaction. No news really is good news. Others might mean distance, signaling that the energy in the relationship has flatlined. The smartest thing you can do when the algorithm says, *"low activity"* is to reach out personally and say, *"We noticed things have been quiet. How are you doing?"*

That small question converts a digital metric into human reassurance. It's the kind of attentiveness that clients notice precisely because it feels optional.

The moral of signals and sentiment

Empathy at scale doesn't come from perfect prediction; it comes from timely awareness. Signals can give you the *when,* but sentiment will give you the *why.* Combine them, and you gain enough foresight to transform maintenance into momentum.

A firm that listens this way never sounds needy or mechanical. It sounds attentive. Technology doesn't create that voice, but it supports it. Think of it like an in-ear monitor that helps a musician hear herself clearly enough to stay in tune.

Once your firm can read the client's signals fluently, the next challenge is choosing how to respond without losing authenticity. In the next section we'll explore how intelligent systems can handle the small gestures of communication—the reminders, thank-you's, and follow-ups—while keeping every exchange rooted in genuine care rather than mass production.

***

Automations that serve, not sell

If "signals and sentiment" teach us how to listen, automation determines how we answer.

In every firm there's an invisible tension between presence and pressure. It's the desire to stay close to clients without crossing into pushiness. How often have we heard the lament from seller-doers that, *"I don't want to be a pushy salesperson."* AI-powered tools now make it possible to send messages faster than we can think, which is both amazing and dangerous. When everything *can* be automated, we need to identify those tasks that are best left to be done manually.

The goal isn't to sound efficient; it's to sound attentive. That's the difference between an automation that serves, and one that sells. Service automations reduce friction for the client. Selling automations reduce friction for you. One creates room for trust; the other crowds it out.

Why we automate in the first place

If we're honest, most automations begin as self-preservation. The project manager is buried, the client list won't stop expanding, and every tool promises that you'll be able to "set it and forget it." Used correctly, automation does buy back time. But only to the extent that we reinvest that time into person-to-person contact.

A thoughtful automation doesn't speak for you; it clears space so you can speak to someone. Every repetitive template, instant reminder, or AI-created summary should ultimately funnel energy toward direct interaction, not away from it.

Ask one quick question before you implement any "time-saver": *"Will this give clients more confidence in us, or merely more content from us?"* If you can't answer yes to the first, skip it.

Signals of the wrong kind of automation

Clients can feel a mechanized message almost immediately. The timing is too flawless, the language too balanced, and the empathy unnaturally even. They may not know a machine wrote it, but they'll sense that there are no human feelings behind it.

Watch for these warning signs:

- The message sends on time but out of sync. It lands when your client is in crisis and it feels tone-deaf.
- The language reads as polished but personless, like a press release.
- Every client receives the same phrase, which is a sure sign that recognition has turned into broadcast.
- The automation continues when silence would have been the wiser option.

Smart firms treat those lapses not as failures of technology but as lapses in judgment. The fix is never more automation; it's more empathy upstream. It's someone pausing to ask, *"Is this the right message for this moment?"*

Designing automation that feels human

The healthiest automations act like stagehands, not performers. They make the show smoother but never step into the spotlight. Here's a quick design test to keep that balance straight:

1. **Automate location, not emotion.**
   Let the system handle *where* something goes—the list, the calendar, the follow-up queue—but let people handle *how* it sounds.

2. **Use technology for timing, humans for tone.**
   An AI scheduler can choose 9:00 a.m. Wednesday based on open rates; only you can choose the line that proves you care.

3. **Send less, personalize more.**
   Re-use structure, never sentiment. Automation should generate a blank space labeled *"<something specific about them>,"* which a person fills before hitting "send.

4. **Turn templated replies into two-step messages.**
   First, a machine acknowledges receipt *("Got it, we're reviewing this!")*. Second, within a few hours, a human adds context *("I*

showed your note to Maya; she's already drafting options.") That cadence preserves empathy while keeping pace.

5. **Set "off" switches.**
   No automation should run indefinitely. Every workflow deserves regular review to confirm it still represents your firm's tone and relationships.

When these principles operate together, the firm starts sounding fast *and* considerate. That's a rare combination that clients instinctively trust.

A quick story: When efficiency spoke too loudly

A regional engineering firm adopted an "AI follow-up assistant" meant to send a personalized note forty-eight hours after every proposal submittal. The wording was elegant, and polished by generative text that can write more elegantly than you or I ever could.

In its first month, two clients wrote back within minutes of each other. One thanked the firm for its attention. The other said, *"Please remove me from your list—this feels like marketing."*

Reviewing the logs, the team saw why. For the second client, the AI had sent the same *"We appreciate the opportunity"* message the morning after they had delivered some bad budget news. Automation had done what automation does. It ignored the context.

The firm immediately changed course. Now, proposals trigger a human review before any follow-up is released. The AI still drafts, but a person scans for mood. The result is fewer messages, better moods, and no more unsubscribes.

The lesson wasn't anti-technology; it was anti-indifference.

Reducing the mundane, not the meaning

Automation's highest calling is to erase repetition, not recognition. A system that drafts project updates, schedules reminders, or pre-loads agendas is doing its job. A system that pushes empathy out of the frame is not. In practice, that means automating *below the waterline*, with tasks the client never sees. And performing *above* it with the calm, personal steadiness that only people can provide.

When your team arrives prepared because the assistant already captured meeting notes and checked roles, everyone appears more gracious, less hurried. That's how clients experience *smart* automation: not as output, but as presence.

The partnership equation

When clients describe their best service experiences, they rarely highlight convenience alone. Instead, they talk about *ease*, and the sense that everything had already been thought of. That "already thought of" feeling is where your marketing and business development automation should aim.

- Automate preparation so your people can be spontaneous.
- Automate the routine so your attention can feel rare.
- Automate reminders so gratitude never feels scripted.
- The result is a rhythm that feels personal at any scale.

The next step in smart client care is teaching both your system and your team to follow through after automation begins. In the next section we'll look at how firms can capture what happens after the handshake. How they can use feedback to reinforce trust and turn performance data into gratitude, not just metrics.

***

Closing the loop after delivery

You know that projects don't really end when the ribbon is cut, or the invoice is closed. The last deliverable isn't a drawing, a report, or a building. It's the feeling the client walks away with.

Smart client care means staying present during that sometimes-awkward hand-off between "done" and "gone." In those few weeks after delivery, clients are forming the story they'll remember and tell about the entire experience. AI can help you listen to that story, but only your human curiosity can make use of it.

The post-project echo

Every completed engagement generates an echo that is heard in small feedback notes, the timing of payment, response tone, their willingness to provide feedback or a reference, or even silence. Those signals are early indicators of whether trust was reinforced or simply maintained out of habit.

Most firms capture technical close-out data centered around time, cost, errors, change orders and the other vital statistics of a project. But they rarely seek out the *emotional close-out.* So, it's vital that you ask the questions that your spreadsheets rarely do:

- *"Was there any moment you felt especially seen or supported?"*
- *"What part of the experience felt unexpectedly difficult?"*
- *"If we work together again, what would you want us to remember?"*

The answers don't belong in a marketing quote sheet first. They belong in the relationship file because they teach empathy as reliably as they teach process.

Feedback as fuel, not a formality

Traditional "client satisfaction surveys" feel like obligatory paperwork. By the time they arrive, the emotion has evaporated, and the checkbox answers are too neat to be useful.

Instead, feedback should be treated as data from an ongoing conversation. Short, open exchanges that are gathered in or close to the moment. Think of client feedback as those interim report cards you used to get in school. They gave you input on how you were doing while there was still time to make important changes toward improvement. By keeping and analyzing transcripts, post-meeting notes, and quick voice memos, you retain the tone and context that text or check boxes on a survey can never capture.

AI can help summarize those inputs without sterilizing them. Imagine a system that reads twenty debrief transcripts and surfaces patterns like:

- *"Clients consistently mentioned feeling calm during coordination."*
- *"Budget communication caused anxiety in multiple projects."*
- *"Several teams praised how issues were flagged early."*

Those aren't scores. They're invaluable anecdotes from the heat of the moment. And they guide the next generation of client experiences more effectively than percentages ever could.

Closing the loop means showing the loop

Feedback only creates trust when clients can see that it changes something. When your firm acts on the input, show it back to the client, quickly and visibly.

- Post a short internal note: *"After the Jefferson Street project, we're revising our kickoff checklist to include a five-minute communications alignment."*
- Send a quick thank you: *"You mentioned that our post-delivery follow-through felt rushed, so we've assigned a single contact to check in with you next quarter."*

Capture these lessons in your knowledge system so the insight isn't lost forever in one desktop folder. That visibility signals professionalism *and* respect. Clients realize their commentary wasn't absorbed by a form; it entered the bloodstream of improvement.

Turning data into gratitude

AI can also help recognize patterns in appreciation as well as complaint. By analyzing messages or testimonials, a system might learn what types of gestures, such as early updates, surprise thank-you's, or recognition of milestones, spark the most positive responses.

Again, this isn't about manipulating sentiment; it's about learning what generates genuine gratitude and what kinds of gestures make clients feel valued and seen. For example, knowing that ninety percent of praised interactions mention *responsiveness*, lets you know where your firm's differentiation lies. You can then celebrate and reinforce that strength intentionally, not accidentally.

The loop closes first with awareness, then with appreciation on both sides.

Human habits make it stick

Automation can collect feedback, but it takes real habits to make it meaningful. That's why many enduring firms build a simple rhythm into project delivery and why AI can help you do the same.

1. *Pause:* Schedule a short, real conversation within 30 days of completion.
2. *Reflect*: With two questions: "What did we learn?" and "What did they feel?"
3. *Record*: Capture insights where both AI and people can find them.
4. *Return*: Share one visible change that grew out of that learning.

That rhythm does more than preserve lessons, it trains culture. Every cycle teaches your people that care is not a single event but a reflex.

When the loop stays open

It's been true in this business since drafting boards and rotary-dial phones dominated the office: True continuity means clients don't experience a gap between projects. Instead, they experience a relationship that simply shifts its focus to the next joint undertaking.

The same emotional metadata that guided you during design can now guide your follow-up. A note such as *"Client sounded proud during ribbon-cutting"* becomes a trigger for later outreach: *"It's been a year since your opening. How has that sense of pride held up?"*

By watching how sentiment shifts over time, AI can highlight when a client's satisfaction starts slipping or when a new concern begins to surface. Noticing those small turns early is what transforms a routine check-in into the chance for another meaningful engagement and the reinforcing of the relationship.

The real measure of closure

The best evidence of smart client care isn't a Net Promoter Score (the metric that tracks how likely someone is to recommend you) or even a repeat contract. It's how easy the next conversation feels.

When a client responds quickly, speaks freely, and assumes you already remember their quirks, the loop worked. Technology helps you hold onto the memory, but empathy is what keeps it meaningful.

The next frontier goes beyond feedback and into predictive trust, which is understanding those behaviors, words, and moments that most reliably strengthen long-term relationships. Before we go there, it's worth pausing on how that trust is protected. Smart client care still depends on clear guardrails that protect privacy, permission, and transparency in how client data is used.

Guardrails for trust

Before any firm can scale empathy, it has to show that the empathy is safe. The smarter our client-care systems become, the more invisible data they hold including notes, moods, fragments of tone, even pauses in email cadence. Used responsibly, that information helps strengthen relationships. Used carelessly, it can make clients feel dissected and studied rather than understood.

That's why smart client care needs the same clarity we expect from good design: well-marked boundaries, visible structure, and honest purpose. The goal is not just privacy compliance. It's emotional credibility.

Transparency builds confidence

Start with openness. Clients shouldn't have to guess what your system knows or how it learns. A simple statement early in an engagement can go a long way: *"We keep brief notes about tone and context, such as when updates feel rushed or when timing seems sensitive, so our team can communicate thoughtfully."*

That line tells them you're attentive, but not intrusive. Transparency shifts the tone from surveillance to service. When people know *why* you're collecting information, they're more likely to offer it freely. And they're more forgiving when technology occasionally misreads the moment.

Permission before pattern

Just because data exists doesn't mean it's yours to analyze. Smart firms make consent part of the relationship, not a buried checkbox. That might mean clarifying which client communications feed into learning systems and which remain private.

A good rule is that you should treat every piece of client data as borrowed, not owned. Borrowed information is handled with care, returned in better condition, and never repurposed without approval.

Protect the human context

AI can summarize every email in a portfolio, but context is still human property. Sensitive exchanges about budget concerns, personnel changes, or internal politics for example, should stay behind clear privacy walls.

If you're training models on client conversations, anonymize names and projects first. Let the patterns do the teaching without exposing the people. This isn't about limiting progress. It's about preserving the trust that fuels it.

Keep humans accountable

Automated empathy still requires human judgment. Every firm should designate someone to review how data is stored, shared, and interpreted. It's best if this is a client-care lead, rather than an IT admin.

When empathy becomes measurable, someone needs to decide where the line between insight and intrusion sits. Think of it as quality control for compassion. The purpose isn't to slow automation down but to keep its aim true.

Trust as the real asset

When a client confides a worry, they've handed you something more valuable than a project. They've given you their context. Your willingness and ability to safeguard that confidence is what makes you and your AI systems credible.

Trust creates permission, and permission creates openness. Without it, even the best technology will only yield polite silence. Smart client care ultimately depends on restraint as much as reach. Privacy, permission, and transparency aren't restraints on empathy; they are its proof.

So, before we look at predictive trust, which is the next step beyond feedback, it's worth grounding the idea in daily practice. How do you make sure these habits of attentiveness and restraint show up every week, not just in policy documents? The answer is a short routine that keeps relationships healthy in real time.

***

The 5-minute *Client Pulse Check*

Smart client care depends less on dashboards and more on steady mindfulness. To keep that awareness active, smart firms adopt a quick weekly rhythm in the form of a five-minute exercise that helps teams stay emotionally current with their important clients. It's not another meeting; it's reflection in action. The goal is to notice patterns early enough to respond with empathy instead of urgency.

Step 1: Pause before the week begins

Once a week, open your client list or dashboard and glance at the few names that matter most right now. You're scanning for emotion, not data. Ask yourself: *"Which of these relationships feels clear and calm? Which feels tense or quiet?"* That intuitive gut check is the human half of the pulse.

Then shift to the digital half and let your AI assistant gather the past week's correspondence, meetings, and small updates. The combination of both views, your instinct plus the system's recall, shows you both the story you remember *and* the one you might have missed.

Step 2: Review the latest signals

Now, read what the system surfaced from the week's emails, meeting summaries, and call notes.

As we discussed earlier in *Context Before Correlation*, you've already built the habit of capturing emotional tone along with facts: those small cues such as *"relieved by cost-savings idea"* or *"cautious about schedule."* This is the moment when that discipline pays off.

Glance through those emotional notes from both people and AI-summarized insights while keeping in mind that you're not auditing the data, you're sensing its mood. Ask yourself:

- *Has their responsiveness changed?*
- *Does the language sound more rushed, cautious, or upbeat?*
- *Are replies shorter or slower?*
- *Is the silence comfortable or concerning?*

If you've set your AI assistant up to tag tone automatically as positive, neutral, or stressed, use that as your starting point. If not, read the correspondence yourself and jot a quick observation. Either way, record it in plain language:

"Tone was lighter after community meeting"

"More abrupt since budget freeze"

These short, human-phrased entries keep emotional context visible across the team. They tell your AI what empathy looks like, and they remind your people that every touchpoint carries a mood as well as a message.

Step 3: Name one thing that's changed

In any ongoing relationship, circumstances are always shifting. It could be a deadline, a changing political wind, or a stakeholder's energy. Name the one change that matters most this week. It anchors the pulse check in reality, rather than rumor. This single note becomes next week's reference point and the heartbeat of the client file.

Step 4: Decide next tone, not next task

Tasks are usually easy and obvious. But tone is strategic. If a client's energy feels anxious, your next move should calm rather than impress. If they feel buoyant, your outreach can carry more momentum. The question isn't *"What do we send?"* It's *"What* feeling *do we want to restore?"* A five-minute tone calibration before action prevents days of avoidable friction later.

Step 5: Close the loop

Before ending the check, capture a short note for yourself or the team:

- *One observation about tone.*
- *One action that matches it.*
- *One appreciation to express when the moment allows.*

That's all. It's five minutes of attention that re-grounds your CRM in care instead of compliance.

Why it works

The 5-minute *Client Pulse Check* links all the concepts in this chapter:

- The empathy habits you formed in *Context before correlation*
- The awareness built through *Signals and sentiment*

- The restraint and authenticity you practiced in *Automations that serve, not sell*
- And the responsibility defined in *Guardrails for trust.*

Together, they translate empathy from a value into a procedure that's short, repeatable, and visible. When your AI tools and your people share that rhythm, the entire firm starts to sound like one attentive voice.

Smart client care doesn't happen once a quarter. It happens in five mindful and repeatable minutes that remind the client, and, just as importantly, you, that someone is still paying attention.

With those five minutes acting as your weekly maintenance loop, your relationships stay alive instead of archived. In the next (and final) chapter, we'll explore how those same cues and emotional breadcrumbs can compound over time. We'll learn how tone becomes trend, memory becomes foresight, and trust becomes something you can actually predict.

Because, as has always been true, this business is based on relationships. Every one of your competitors can do good—even great—work. That's not where you gain competitive advantage.

But, when you integrate the tools we've been talking about, the advantage you gain is powerful and invaluable.

CHAPTER 18
Your Firm's Next [re]wire

Re-charging the circuit

Every firm heads into a change like this with energy. New ideas, new tools, new results. And then, over time, that energy eases back into routine. Things still work, but the pulse that kept everyone curious starts to fade a little. That's normal. Even the best systems and habits begin to coast after the excitement wears off.

The goal now isn't to chase that first spark again. It's to keep a steady current running through the work. That means checking in occasionally, seeing what's become automatic, and deciding what still deserves attention. Innovation runs on small adjustments, not dramatic overhauls.

Think of it like keeping equipment tuned. You don't rebuild the machine every season. But you listen for new sounds, tighten a few bolts, and clean the filters. The same is true here. The [re]wire you began when you brought AI and new thinking into the firm will only stay strong if you give it regular care.

That care shows up in simple ways such as asking what's working, noticing what's not, or sharing one short lesson after a project wraps instead of letting it fade. Those small habits do more for long-term momentum than any new software version or big training session.

So, before we talk about the next plan or checklist, let's pause for a moment. Remember why you started this in the first place. You wanted to make work clearer, easier, more rewarding for the people doing it, and more responsive to your clients. Keeping that spirit alive is the real maintenance. Everything that follows in this chapter is just a way to help you do that on purpose.

The 90-day [re]wire plan

When the energy around the changes you've made starts to fade, the best way to bring it back isn't with another big project. It's with a rhythm that keeps the team learning and adjusting in small, steady ways. Ninety days is long enough to notice results but short enough to stay within reach. Think of it as one healthy quarter—a routine that helps momentum find its footing again.

The plan has four parts: *assess, pilot, adopt*, and *share*. The key is to keep it light. You're not writing a whole new strategic plan. You're building and sustaining a rhythm.

1. Assess: Where you stand

Start by taking a look at how the last few months have gone. What's actually helping? What got set aside? Which tools or habits feel heavy again?

Invite small, honest answers. You don't need a full audit; just a few clear observations from the people using the tools every day are enough. The goal is to spot drift early before it turns into frustration.

Questions to ask can include:

- *Where have we seen the fastest improvements?*
- *Which tasks still feel harder than they should?*
- *What have we learned that's worth repeating?*

2. Pilot: Try one new thing at a time

Pick one area that matters and run a short trial. Maybe it's reshaping how you debrief projects, testing a new AI summary tool, or checking how client feedback gets shared. Keep the scale small and the experiment safe.

Write down what "good" would look like before you start, otherwise success will be impossible to spot. When the pilot finishes, decide together whether it's worth turning it into standard practice or letting it go.

3. Adopt: Make what works visible

If something clearly helps, move it into the regular workflow right away. Document the small details: what prompt, what step, what habit made the difference? This will allow others to use it without guessing.

Don't turn every new success into a big policy announcement. That slows the system. A quick note on your shared drive, a five-minute staff huddle, or a short Teams message saying *"Here's what worked and why"* is plenty.

4. Share: Turn learning into culture

At the end of each ninety-day cycle, hold a single meeting to talk about what changed. Highlight one improvement, name one lesson, and agree on one next step. Keep it conversational. Let people see progress in plain numbers or stories. This is less about "initiatives," and more about locking in new "habits."

Sharing closes the loop. It reminds everyone that they, and the firm, are learning in real time, not waiting for a big annual review to catch up.

A 90-day [re]wire isn't meant to add pressure. It's a way to keep motion steady and human. It turns "innovation" into something you can manage with the same ease you manage projects. It has clear intent, steady communication, and time set aside to check if things still make sense.

Run that cycle a few times, and renewal becomes normal. Because you're not starting over each quarter. You're just keeping the current strong.

Measuring momentum without tension

At some point, someone will ask, *"How do we know this is working?"* It's a fair question, and the right one. But how you answer it matters.

If you respond with dashboards, bar charts, or a dozen metrics, most people tune out. The goal isn't to count everything. It's to notice whether the work is getting easier, clearer, and more connected and to confirm that your clients are being served in a more connected and empathetic way. Progress in a modern firm doesn't always show up as bigger numbers. More often, it shows up as steadier confidence.

Start with simple signals:

- *Rhythm:* Are meetings shorter because people already know what to do?
- *Response:* Are clients getting answers faster and with fewer drafts?
- *Relief:* Does the team seem less frantic when deadlines hit?

It's not unlike setting a goal to become more fit. You can weigh yourself, measure your BMI, and set up charts to meticulously track calories. Or you can simply ask yourself if you're getting out to exercise every day and if you're eating smarter than you were before. Those things don't fit neatly on a graph, but they're real measures of momentum. They tell you that learning is working its way into habit.

You can still track a few small numbers, just be thoughtful about which ones. Look for *time regained,* not hours billed. *Errors caught early,* not work pushed faster. *Participation in learning loops,* not just completion rates. Metrics should confirm progress, not create anxiety.

Keep evaluation open and conversational. Once a quarter, ask three quick questions:

1. *What feels easier now than it did 90 days ago?*
2. *Where are we still guessing or duplicating effort?*
3. *What story could we share that shows this change helping someone—inside or outside the firm?*

Anecdotes and stories may not seem like data, but people trust what they recognize. A two-minute story in a staff meeting about a smoother proposal or a calmer client call will persuade far more effectively than a slide full of percentages.

If a conversation about progress feels joyful instead of judgmental, you're measuring the right things. Momentum should give people confidence, not tension. The moment tracking progress starts to feel like grading performance, stop and reset. You're not evaluating people. You're taking the pulse of a system that's finding its rhythm again.

In the long run, the truest measure of success is calm. When teams know what matters, use their tools comfortably, and still sound like themselves, the rewiring has taken hold.

Updating your Human + Machine balance

When you first began this journey, it was easy to think of "human" and "machine" as opposites that occupy separate lanes on the same highway. Early on, you probably spent time debating what should stay in the hands of your people and what could safely shift to AI. Those conversations were helpful, but temporary. The truth is, that line keeps moving, and it always will.

Every new capability—an upgraded CRM, a smarter predictive tool, a quicker language model—nudges the balance. The key is not to freeze the boundary, but to revisit it calmly every so often. The healthiest firms make this a simple checkpoint, not a philosophical debate.

Start with two lists

1. *What tasks are making us smarter, faster, or more consistent because of technology?*
2. *Where do we still need a person's voice, empathy, or judgment to make the outcome feel right?*

You'll probably find that the best results happen where the lists overlap, in those instances when the facts are handled by the machine and the meaning is clarified by a person.

For example, an AI assistant might summarize a client's feedback in seconds. But it still takes a thoughtful project manager to hear what the client meant between the lines. Only a person can interpret the pauses, the hesitations, and the tone that never shows up in the text. That combination is the balance you're trying to maintain: efficiency paired with understanding.

Re-checking that balance every few months keeps both sides honest. If the tools start to feel like they're steering the conversation, pull them back a little. If people are spending too much time on manual routines the system can manage, adjust the other way. The right mix will have a lightness to its feel. Technology will be carrying the load while people continue to guide the direction.

Make the review simple. Once or twice a year, ask in a team meeting:

- *Where did technology genuinely save us time this quarter?*
- *Where did it create more noise than value?*
- *Did we still sound like ourselves in the messages and materials it helped produce?*

You're not looking for perfect ratios; you're looking for comfort. The goal isn't to have more automation, it's to have more human moments because of it. And when you have the balance right, the tools disappear into the background, and what clients notice is your attention, not your technology.

If your people feel trusted and your clients feel remembered, you're in the right place on the curve.

Obstacles and course corrections

Even with the best systems and intentions, it's easy to drift. Every firm does it. Deadlines pull people back into old habits, and good ideas lose ground to the next fire drill. It's not a failure of discipline; it's just how real work behaves when time and energy get thin.

That's why gentle course correction matters more than constant control. You don't need grand resets or new initiatives every time momentum dips. Like driving a car, it's far easier to make tiny, regular corrections through the steering wheel, than to try recovering from a massive skid. You just need the habit of noticing small changes early and bringing things back in line before they turn into chronic problems.

Most of the warning signs are easy to spot:

- *Too many tools.* If people have to open five screens to finish one task, your helpful technology has become cluttered. Simplify. Decide which systems actually earn their keep and combine or retire the rest.
- *Uneven tone.* If proposals or client messages start sounding mechanical again, pause. Let a few human voices rewrite key sections to reset tone before automation spreads the drift. Remind your AI of what you want and how you want it to sound.
- *Fatigue dressed as efficiency.* If staff are completing more tasks but smiling less, you're measuring motion, not improvement. Slow things down long enough to ask, *"Is this still working for us?"*

When any of those signals show up, call a short "reset conversation." Ten minutes is usually enough. Ask what feels off, what changed, and what a small fix would help. The tone should be curious, not corrective. You're treating the system, not scolding the people.

Remember that AI-assisted work exaggerates both good and bad habits. A small misalignment in process or culture can scale quickly once automation joins in. Don't wait for a quarterly review to catch it. The first person who notices friction has permission to raise it, whether that's a principal or an intern. Correction lives best close to the work.

Most detours are minor. A prompt that needs updating. A checklist that grew too long. A meeting that turned into a monologue. Each one takes just a few minutes to steer back. Over time, those little realignments keep the bigger picture steady.

So, when you catch the firm drifting, don't announce a recovery campaign. Just ease the wheel. Remind people that course correction is part of progress. Consistent curiosity beats constant reinvention.

The measure of a mature culture isn't flawless execution. It's the ease with which small mistakes are noticed and fixed. And if each correction leaves the team a little calmer and a little wiser, then the system's working exactly as it should.

Celebrating progress

When a firm starts to run smoothly again, it's easy to skip celebration and move straight to the next project. Professionals tend to do that. Early in my career, I worked for an architect who told us loudly and emphatically that, as professionals, we ought to know when we'd done something right. His job was to point out when we'd done it wrong.

But recognition is the fuel that keeps improvement alive. People need to feel that the effort mattered and that their adjustments added up to something real.

Celebrating progress doesn't have to be loud or staged. In fact, the most meaningful recognition usually sounds like conversation, not ceremony. It's a few quick examples shared over coffee, a principal who thanks a coordinator for refining an AI prompt that saved the team an hour, or a project manager who points out how the new meeting rhythm kept a client calm. These small acknowledgments build pride because the motivation comes from meaning, not applause.

Try marking the end of each 90-day cycle with a short reflection. Keep it comfortable and human by asking:

- *What felt smoother this time than last?*
- *Who made a small change that helped everyone else?*
- *What will we carry forward into the next round?*

Write the answers on a whiteboard or drop them in chat where everyone can see. You don't need trophies or slogans, just clarity about what's working and gratitude for who made it work. The act of naming progress reminds people they're part of something moving forward.

Storytelling helps too. When you retell a success, even a small one, you reinforce shared confidence. *"Remember when the client mentioned how quickly we turned that proposal around—that's because the AI draft saved us half a day, and the team used that time to personalize the closing section."* Those details matter. Stories like that have more impact inside a firm than any motivational poster ever could.

And remember, celebration isn't just for wins. It's also for persistence. Make sure you recognize the times when the team stayed thoughtful under pressure, corrected a mistake quickly, or kept empathy alive during a tough week. Those are the moments that define culture far more than perfect metrics do.

When recognition becomes part of your rhythm, people start looking for what's going right, not just what's going wrong. That shift of attention changes everything. Optimism becomes a habit, and curiosity feels safe again. The pulse returns without fanfare.

The best proof that your [re]wire is holding isn't a graph or a policy; it's the obvious confidence of a team that enjoys the way it works together. When you see that, pause long enough to notice it out loud. Then keep the current moving.

Try this now: The *Quarterly [re]wire Checklist*

Once each quarter, gather the people who keep the firm running including the principals, project managers, marketing staff, and the support team who make sure all the pieces connect. Set aside thirty minutes. No slide deck, no big meeting title. Just a simple conversation about how the AI-related systems, tools, and habits are holding up.

Use this short checklist to guide the discussion. You can adjust it to fit the size and rhythm of your firm, but it works best when you keep it conversational and honest.

Quarterly [re]wire Checklist

1. **Energy and focus**
    - Where do we still feel momentum?
    - Where are people slipping back into old habits?
    - What's one process that could move from "necessary" to "simpler"?
2. **Human + Machine balance**
    - What work felt lighter because of AI or automation?
    - Where did the tools create more friction than flow?
    - Did our tone, proposals, and communication still sound like *us*?
3. **Learning and improvement**
    - What did the team learn about our clients, our tools, or ourselves that we should record and teach forward?
    - Were any small frustrations fixed quietly that deserve to be shared?
    - What pattern keeps surfacing that might need a bigger decision next time?
4. **Client experience**
    - Did anyone notice faster responses, fewer revisions, or clearer messaging from us?
    - What piece of feedback stood out as proof that we're easier to work with?
    - Where did we miss a chance to reassure or surprise a client in a good way?
5. **Culture and well-being**
    - Does the pace feel sustainable?
    - Are we still finding time to talk about work like people, not just through dashboards?
    - Who could use a little acknowledgment right now?

Keep the notes light with bullets instead of paragraphs. Summarize what you hear into two lists: *"Keep Doing"* and *"Adjust."* That single page becomes your guide for the next quarter.

Before closing the conversation, ask one final question: *"If we did only one thing differently next quarter to make work feel clearer and more human, what would it be?"*

There's no need for a formal action plan. The value is in having the conversation regularly, not perfectly. Done four times a year, it keeps your leadership, your tools, and your culture aligned without waiting for a crisis to remind you.

A short talk like this costs half an hour and buys months of clarity. That's momentum in its simplest form: *a firm that listens to itself before the marketplace brutally and impartially reminds it to pay attention.*

By now, you've seen that [re]wiring a firm with AI isn't about chasing the next platform or polishing the latest prompts. It's about maintaining a culture that knows how to learn together, consistently, and without panic.

The tools you've added will keep changing. New models will appear. Clients will ask harder questions about speed, pricing, and originality. Through it all, the part that still decides your future isn't the algorithm. It's the judgment, empathy, and steadiness of the people using it.

Technology can, and will, widen your capacity. But only people can choose what to care about. Every confident response, every thoughtful paragraph, every moment of trust that wins a client will all come from the human side of the equation.

As you move into the closing section of this book, think of it not as an ending but as a return. We've talked about structures, processes, and tools. Now we come back to the one element they all serve: The [re]wire is the means. The human is the reason.

EPILOGUE

Why the people still win

Looking back to the first [re]wire

A decade ago, when the first *[re]wired* book appeared, the world of marketing for architects and engineers still moved at an analog pace. Firms were proud of their portfolios but uneasy with their pipelines. CRM systems sat half-empty, websites looked interchangeable, and "business development" often meant whoever had time left on Friday afternoon.

That first *[re]wire* wasn't about technology so much as mindset. I asked leaders to stop mistaking activity for strategy and, instead, to build deliberate marketing systems that could be measured, managed, and improved. It was the start of digital literacy for a profession that had long relied on word-of-mouth and reputation to do the heavy lifting.

And it worked. Slowly, across the industry, firms learned to treat pursuit planning and client intelligence as disciplines, not mysteries. Proposals gained structure. Conversations gained purpose. Business development began to sound less like chasing opportunity and more like understanding need. We stopped confusing chance with choice.

The courage we needed then was the willingness to question what "good enough" really meant. We learned to use data as a mirror, not a verdict, to see where our assumptions were thin and where our strengths actually lived. That shift created the blueprint for how modern AE firms sell: insight first, instinct second.

That first rewiring didn't make us depend on technology; it made us depend on clarity. It taught us to think like marketers and act like partners. And in doing so, it quietly—perhaps inadvertently—prepared us for the next leap. That's the one that's arriving now, faster and louder than any we've faced before.

This moment: The second [re]wiring

Artificial intelligence has arrived with the kind of scale and speed that rewrites not just the rules of work, but the rhythm of thought. Overnight, the tools we once dreamed about are sitting patiently on our desks, waiting to draft, format, summarize, and analyze at a pace no human can match.

To some, that's exhilarating. To others, unsettling. And to most, it's both. But whether welcomed or resisted, this second [re]wire is already here.

Just like the first transformation, it's not really about the machine. It's about how we learn to *think* again. When AI lifts the weight of routine, it exposes what's left: judgment, empathy, and imagination. Those are the new frontiers of professional advantage. The technology will do what it does best—pattern, predict, and produce. But the meaning of those patterns still belongs to us.

In the business of selling design services, that distinction matters. Our clients don't come to us for syntax or speed; they come for confidence. They want to trust that someone sees the bigger picture and can guide them toward a solution that feels both wise and safe. AI can process the data; only a human can interpret the stakes.

So, our task in this next [re]wire isn't to compete with artificial intelligence, it's to pair with it. To use the glow of technology not as a spotlight on us, but as a lantern that makes our clients' needs easier to see. That's how credibility is earned now: through precision and presence working side by side.

Firms that thrive in this era will be the ones that understand the rhythm of collaboration between human and machine. They'll treat AI as an accelerant, not a driver. As something that sharpens response and frees time for curiosity. Done well, that partnership feels effortless. The client notices the clarity but never the code.

This second [re]wiring demands the same trait the first one did: courage. Not the adrenaline of disruption, but the quieter courage to keep learning. To stay curious enough to ask not, *"What can this tool do?"* but, *"What should we do with it?"*

That simple, and very human question will decide who leads the next decade of selling in professional design services.

What doesn't change

Every generation of tools promises to revolutionize selling. But the heart of the work—the human conversation between a professional and a client deciding whether to trust one another—has never changed. Technology can accelerate that moment, but it can't replace it.

Relationships anchored in credibility and empathy still win. Clients still want evidence that you understand them, that you see not just their project but their pressure points, their ambitions, and their risk. No algorithm can look a client in the eye and say, *"We'll take care of this."* That assurance still belongs to people.

AI can write, but people must interpret. It can predict patterns, yet only people can sense intent. Language models string words together, but they don't know which words foster trust. That knowledge, which can only be born of listening, empathy, and judgment, is still entirely human.

Data can illuminate and logic can persuade. But belief, the gut conviction that your firm is the right partner, only comes when a human voice confirms it. As clients adopt their own AI tools, that difference will become even clearer. They won't be impressed by your technology—they'll be reassured by your relevance.

The professional's job now is to join the two: *insight plus intent.*

- Insight comes from the machine with its pattern recognition, context, and precision.
- Intent comes from the person with their curiosity, credibility, and care.

When those two meet, great marketing and selling can happen. The conversation feels effortless, but underneath sits a web of awareness in which smart tools feed sharper questions, and sharper questions deepen trust. That rhythm, the combined power of machine speed with human sense, is the real [re]wiring.

In the end, the tools have changed, but the truth hasn't. The client is still the center of the universe. Every new system we adopt only matters to the extent that it helps us orbit them more closely, listen more carefully, and serve more meaningfully.

Your firm's new voice

If the first [re]wiring taught us to systemize our marketing and the second is teaching us to humanize our technology, the next horizon is learning to *speak with a new voice*—one that sounds unmistakably human yet is powered by machine intelligence.

Over the coming decade, the most effective firms will not sound busier, they'll sound *clearer.* With automation easily handling coordination, formatting, and data gathering, the human voice will no longer have to shout over noise. Messages will read simpler, proposals will feel more conversational, and client communication will carry the calm authority of people who finally have time to think.

That clarity will become a competitive advantage. Clients, just like the rest of us, are weary from filtering noise. They navigate inboxes full of synthetic language including proposals that read like templates and emails stuffed with keywords. When they encounter a message that sounds real, it's disarming in the best way. They pause and they listen.

AI makes this possible, not by writing for us, but by giving us the time and insight to write *to* someone. It allows firms to move from volume to value. From sending more information to making more sense. The technology behind the scenes is impressive, but the result feels entirely human with communication that is concise, relevant, and warm.

Every great brand voice echoes the sound of its culture. And the firms that do this well will discover that tone is not decoration. In fact, and importantly, it's differentiation. Authentic language builds confidence faster than any credential list ever could.

The AE industry has always balanced art and precision. Now it can balance technology and presence in the same way. The tools will handle the scaffolding, including the frameworks, the facts, and the forms. That will leave people to shape the story. AI can suggest the words, but only people can decide which ones truly earn trust.

This is the coming age of *hybrid selling*: part data, part dialogue. The firms that thrive will use AI to remove friction, freeing their teams to spend more time on interpretation, counsel, and relationships. Ironically, technology will let the best firms sound *more* personal than they ever have before.

The signal will once again be stronger than the noise. That's the real rewiring of voice. It will result in clients don't hear a system, but a person who understands them.

The send-off

If this book has done its work, you'll leave not just with tools, but with perspective. Every [re]wire, whether it's digital, cultural, or now AI-driven, has been about clarity. Each time new technology arrives, it gives us another chance to see our clients and ourselves more honestly.

When we strip away the novelty, what remains are the same imperatives that have always defined how design professionals win work: *Listen well, think clearly, and care more than the next firm does.*

Those simple acts are about to matter even more. In an AI-enabled world, sameness has never been easier to produce. Every message is polished, every proposal formatted, and every answer generated. What's scarce now is authenticity. And authenticity is the one thing no system can automate.

So, as you bring these tools into your firm, treat them the way a good engineer treats structure: it's something that supports vision, instead of replacing it. Use AI to give shape and speed to the work, but let human judgment continue to give it meaning. The best firms will be the ones that remember the order of operations: *people lead, machines assist.*

The future of marketing, BD, and sales in the AE world isn't artificial. It's *amplified.* It's the same integrity, just seen through sharper and faster channels. What wins next isn't fluency with technology, it's fluency with people, strengthened by technology that finally serves them instead of stealing their time.

A decade ago, I dared you to be bold. I invited you to stop blending in and to start sounding different. That dare still stands. Only now it comes with better tools and a clearer purpose. Use them to deepen your relevance, sharpen your empathy, and widen your reach.

Because in the end, every advance in intelligence—artificial or otherwise—reminds us of the same truth: progress is mechanical, but trust is human. And the firms that continue to earn trust will always, unmistakably, win.

APPENDICES

Appendix A

Ethical AI use guidelines for professional firms

Why this matters

Professional integrity doesn't end when a tool starts to "think." AI can deliver new speed and scope, but your credibility still depends on the judgment, confidentiality, and care of people. These guidelines will help keep your firm aligned with the same principles that built your reputation in the first place. Use them as a short code of conduct, a teaching aid, or the one-page reminder that sits on the wall beside your project charter.

1. **Purpose before automation**
    - Define intent first. Every AI use must answer: What human purpose is this serving?
    - Avoid automation for its own sake. If the only reason is "because the software can," stop.
    - Link each AI feature to a measurable client, ethical, or internal benefit.

2. **Human ownership never ends**
    - A named person reviews, corrects, and approves every AI-assisted output before it is shared outside the firm.
    - No AI tool holds authority. It provides suggestions only.
    - Communicate internally. The assistant recommends; the person decides.

- When something goes wrong, accountability stays both human and visible.

3. **Transparency builds trust**
    - Disclose AI's role in plain language when its output reaches a client or public audience.
    Example: *"Portions of this document were prepared with AI assistance and reviewed by our team."*
    - Keep disclosure brief, calm, and factual. It's neither apology nor hype.
    - Maintain an internal log of AI-assisted tasks for audit and learning.

4. **Confidentiality by default**
    - Never paste client names, financial data, or proprietary material into open or public models.
    - Use enterprise or locally secured versions with clear non-training assurances.
    - Redact identifying details ("Client A," "regional hospital") in test prompts.
    - Treat every AI environment as public unless proven private in writing.

5. **Data integrity and traceability**
    - Store an original and AI-edited version of any official communication.
    - Verify every fact, citation, and quantitative claim through independent sources.
    - Make it possible to trace what tool was used, when, and by whom.
    - Keep your datasets current. Out-of-date information corrupts trust faster than inaccuracy.

6. **Bias awareness and representation**
    - Assume all datasets contain cultural bias until checked.
    - Read outputs for stereotypes, exclusion, or tone imbalance before publishing.
    - When possible, test drafts across multiple reader perspectives or demographic lenses.
    - Make a note of biased patterns. Use them as teaching data for future prompts.

7. **Informed consent and client respect**
    - Treat all data as borrowed, not owned.
    - When monitoring sentiment or behavior, inform clients what's being observed and why.
    - Use insights for service, not surveillance.
    - Decline any analysis that would surprise the client if explained aloud.

8. **Continuous learning and correction**
    - Establish a short "AI Review" slot in team meetings to surface odd results or ethical questions.
    - Record important discoveries and fixes in an accessible log.
    - Refresh firm-wide guidelines quarterly as technology and laws evolve.
    - Celebrate good catches. They prove that oversight works.

9. **Measured disclosure in public and digital media**
    - If discussing your firm's AI capabilities, emphasize stewardship over novelty.
    - Avoid language that implies replacement of professional expertise.
    - Keep marketing statements consistent with actual practice and data-protection policies.

10. The reputation test

- Before releasing any AI-assisted material, pause for the three-second gut check:
  - Would we be comfortable signing our name to this?
  - Would our client feel respected if they knew how it was made?
  - Would we still be proud if it were shared publicly tomorrow?
- If all three answers are yes, you're operating inside the ethical current that sustains professional trust.

Endnote

While technology continues to change weekly, ethics almost never need rewriting. Use this checklist not as bureaucracy but as ballast. It's the weight that keeps innovation upright. Because in every era, and especially this one, while speed might impress a client once, your integrity will keep them around for the long term.

Appendix B
The prompt library

Why this matters

Prompts are neither magic words nor tricks. They are professional tools that you use to provide leadership and guidance in text form to your AI Assistants. They transform vague intentions into clear direction. In the design professions, where judgment, timing, and nuance can be vital, strong prompts help you keep AI aligned with your firm's ethics, voice, and expertise.

The quality of what AI produces will always reflect the clarity, empathy, and precision of your request. This library offers ready-to-use examples and guiding principles so you can interact with AI as a thinking partner, not a keyboard trick.

Before you simply cut-and-paste the example prompts, read this short introductory section to ensure you're using the prompts to your greatest advantage.

1. **Prompt principles and etiquette**

   Every reliable prompt follows the pattern:
   Context + Action + Outcome = Clarity

Element	What It Means	Example Clause
Context	Who you (AI Assistant) are and for whom you speak.	*"You are proposal-writing expert assisting a 35-person architecture firm responding to a municipal* RFP.*"*

Element	What It Means	Example Clause
Action	The specific task.	*"Rewrite the approach section for clarity."*
Outcome	The desired tone or format.	*"Keep it plain language, respectful, and under 200 words."*
Context	Who you (AI Assistant) are and for whom you speak.	*"You are an engineering lead preparing internal guidance for a multidisciplinary project team beginning schematic design on a large campus renovation."*
Action	The specific task.	*"Summarize the key coordination risks between structural, MEP, and architectural scopes, using bullet points grouped by discipline."*
Outcome	The desired tone or format.	*"Deliver a one-page overview in professional but conversational tone—something colleagues can scan in two minutes during the kickoff meeting."*

Always write your prompts in full paragraphs, as if you were actually explaining your instructions to a colleague. Never give the AI Assistant single sentences or sentence fragments without providing the context in which they are working and the deliverable you want. Providing the context builds its comprehension, not just for the current prompt, but for the larger, ongoing scope in which you're working. Issuing terse commands breeds confusion and guarantees that your results are going to be disappointing.

2. The 3-step rhythm

The likelihood of getting the result that perfectly matches your intent on the first prompt is slim, especially when you're new at working with AI or when you are first beginning a new project with your assistant. As with working alongside human colleagues, it takes a little time and a little back-and-forth before everyone is on the same page.

Use this three-step rhythm to home in on the optimum output:

- **Ask** - Use the *Context* + *Action* + *Outcome* pattern of prompting to obtain the first draft.
- **Adjust** - Review the result, then use plain, conversational language to describe to the Assistant what worked and what didn't.
- **Direct** - Use the same conversational tone to request the next version with sharper focus.

3. Use role and tone to fine tune the fit

You may find that you need to clarify the context, the audience for whom the output is intended, the voice and tone in which you want it written, or some other variable that the Assistant didn't get quite right. By specifying the role you want the Assistant to play, you give it implicit filters and vocabulary, which will make the difference between it using specialized technical jargon or plain language clarity. Good prompting will feel like coaching another human, not coding a computer.

- *Respond as a project manager preparing a one-page brief for a mayor.*
- *Explain as a policy analyst advising a community board.*
- *You are assisting a multidisciplinary design firm preparing a proposal for a city client. Draft a two-paragraph executive summary in a clear, confident tone suitable for non-technical reviewers.*

Prompts by purpose

What follows is a list of 100 different prompts to get your thinking and your conversation started with your AI Assistant. Use this list as a quick-reference library from which you can select, adapt, and then refine.

Mastery comes from use, so don't be afraid to experiment. Recall how, in Chapter 2, I encouraged you to set up a safe, private sandbox for experimentation? In that space you can run small, low-pressure experiments with these prompts to see what results you achieve. Remember to treat the space as practice, not production. It's your personal AI lab.

After you begin to feel more confident, you can begin to adapt these prompts for use in your real, client-facing work. Save each successful prompt in a shared Prompt Library. Compare results, refine collectively, and keep track of the versions as deliberately as you do for your design iterations.

AI doesn't make you faster; clarity does. These prompts are only shortcuts to that clarity.

Important Note:
In the *Context* + *Action* + *Outcome* rhythm I have encouraged you to use, these examples provide the *Action* component only. You must add the context and desired outcome.

Research & Insight

Use when you need perspective, context, or synthesis before drafting anything.
Tone guidance: Analytical and objective.
Safety: Verify sources before re-use; treat results as hypotheses, not facts.

1. "Summarize the last 12 months of public statements from <Client Name>. Identify recurring words that signal strategic priorities."
2. "Map the top five pressures facing municipal infrastructure projects in <year> based on industry publications."
3. "Read this RFP section and list the implicit client fears or risks it reveals."
4. "Create a brief of emerging policy shifts that could affect hospital design funding in the next two years."

5. "Explain new energy-efficiency directives in plain language for a client newsletter."
6. "Compare two competitors' messaging about sustainability and highlight unique positioning gaps we could fill."
7. "Identify trends in urban-mobility planning in Europe and translate their relevance to U.S. mid-size cities."
8. "From this demographic dataset, suggest three community narratives we might emphasize in planning conversations."
9. "Summarize findings from five academic papers on adaptive reuse and list practical implications."
10. "Extract unspoken assumptions behind recent government grants related to infrastructure resilience."
11. "List current funding mechanisms for affordable housing and rank by accessibility for municipalities."
12. "Translate this 40-page feasibility study into five client-facing insights under 100 words each."
13. "Outline legislative milestones that shaped this regulation's timeline."
14. "Review these public hearing transcripts for recurring stakeholder concerns."
15. "Distill this technical guideline into a 200-word explainer for non-engineers."
16. "Summarize media sentiment around smart-city technology deployment in <year>."
17. "Create an annotated timeline of research milestones for net-zero manufacturing."
18. "List common misinterpretations of 'resilience planning' in public discourse and how to clarify them."
19. "Identify local success stories we might cite as credible precedents for this project type."
20. "Highlight three statistics from this dataset most compelling for a lay public presentation."

Writing & Editing

Use when refining tone, clarity, or persuasion.

Tone guidance: Keep outputs professional, confident, and client-focused.

Refinement Tip: Ask "what works/what distracts" after each draft.

1. "Rewrite this proposal section to emphasize outcomes rather than process."
2. "Edit this paragraph for clarity and economy. Remove unnecessary words, prefer strong verbs, and favor plain, direct language."
3. "Recast as a short, benefit-focused headline. Make the message persuasive, concrete, and easy to grasp at a glance."
4. "Translate this technical section for a city-council audience."
5. "Tighten this email into three crisp paragraphs with a friendly-professional tone."
6. "Adjust this narrative to suit a public presentation with more warmth and less formality."
7. "Convert bullet notes into a story arc with problem-solution-impact structure."
8. "Check this draft for repetition, jargon, or clichés and suggest replacements."
9. "Polish tone from defensive to proactive without changing facts."
10. "Rewrite this client letter with empathy and confidence. The goal is reassurance."
11. "Edit for rhythm and reading flow. Vary sentence length and maintain calm authority."
12. "Format for clarity. Add headers and bullet points without changing meaning."
13. "Change perspective from firm-centric ('we') to client-centric ('you')."
14. "Recast this copy in plain language with a reading level of 9th Grade or below."

15. "Draft an executive summary under 150 words that finishes with a strong forward-looking statement."
16. "Review for hidden assumptions or bias in word choice and provide neutral phrasing."
17. Blend concise clarity with professional warmth. Produce two contrasting versions; one plain-spoken and direct, the other polished and executive to compare tone and readability.
18. "Turn this internal memo into a LinkedIn post while preserving confidentiality."
19. "Write a call-to-action paragraph that sounds helpful, not sales-driven."
20. "Adapt this text for use in a slide with short lines and high clarity."
21. "Convert this passive-voice section into active without sounding aggressive."
22. "Summarize the entire report in three sentences suitable for an award submission."
23. "Condense this article into ten key-point bullets for a talking-head video script."
24. "Generate a proofread checklist for this draft based on clarity, empathy, and compliance."
25. "Offer two headline variations, one analytical and one emotional."

Idea Generation & Innovation

Use when stuck or seeking fresh angles.

Tone guidance: Curious and exploratory.

Safety: Mark speculation as such; never present creative drafts as approved facts.

1. "List ten unexpected metaphors for design resilience using everyday objects."
2. "Generate alternative value-proposition statements emphasizing community benefit."

3. "Imagine three post-carbon futures for our industry and how our firm could thrive in each."
4. "Brainstorm partnering opportunities that align with this project's sustainability goals."
5. "Propose analogies connecting infrastructure to human health outcomes."
6. "Invent a client workshop exercise to encourage alignment on design priorities."
7. "List potential unintended consequences of this proposed policy."
8. "Explore counterintuitive advantages of smaller budgets."
9. "Reframe this challenge as three 'How might we…?' questions."
10. "Generate ideas for explaining complex data visually to non-technical stakeholders."
11. "Offer five ways to make project close-out meetings more memorable."
12. "Devise social-media topics connecting civic value with technical achievement."
13. "Suggest creative uses for unbuilt or interim project spaces."
14. "Imagine how AI might assist community engagement without replacing human presence."
15. "List client-education materials we could automate responsibly."
16. "Brainstorm tagline options that balance authority and approachability."
17. "Create three versions of a pledge clients could sign to demonstrate shared responsibility."
18. "Reverse-engineer our competitor's message and reframe it with our strengths."
19. "Draft a two-minute elevator pitch for <year> urban-renewal challenges."
20. "Suggest a new internal learning series about emerging tech ethics."

Project Management & Process

Use when improving efficiency or planning workflow in teams.
Tone guidance: Neutral, factual, and concise.
Safety: Keep internal; avoid entering proprietary project data into public tools.

1. "List the critical path items for this deliverable with milestones and dependencies."
2. "Write a meeting agenda focused on decisions rather than updates."
3. "Suggest clear follow-up notes from these raw meeting transcripts."
4. "Summarize recurring client questions from emails into an FAQ."
5. "Build a table estimating staff hours by phase using historical data."
6. "Draft a risk-management checklist for early design review."
7. "Create talking points for a scope-change meeting. Make it neutral, factual, and concise."
8. "Provide a template for documenting learning after project close-out."
9. "Outline four communication checkpoints for a remote hybrid team."
10. "Develop a five-item dashboard of leading indicators for project health."
11. "Turn these QC notes into an inspection checklist."
12. "Draft polite reminder wording for overdue client inputs."
13. "Suggest process improvements based on this workflow description."
14. "Rephrase technical issues into risk-language suitable for executives."
15. "Develop a succession plan outline for key personnel transitions."

Client Engagement & Business Development

Use when preparing communication that builds or renews relationships.
Tone guidance: Warm, trustworthy, and concise.
Safety: Respect confidentiality and brand voice. Avoid generating copy with personal-data examples.

1. "Draft a thank-you letter after a client presentation. Make it appreciative, brief, and credible."
2. "Compose three intro messages for reconnecting with past clients via LinkedIn."
3. "Write a proposal opener that converts technical specs into benefits."
4. "Outline discovery questions for an initial client meeting."
5. "Summarize win themes across our recent proposals."
6. "Create alternate taglines for this pursuit capturing both trust and innovation."
7. "Draft a project-award news release in a professional, humble tone."
8. "Condense this project description into a 100-word portfolio blurb."
9. "Rephrase our boilerplate to sound more human without losing authority."
10. "List ethical red-flags we should inspect before pursuing new partners."

Reflection & Learning

Use during debriefs or end-of-week reviews to convert experience into lessons.
Tone guidance: Reflective and candid.
Safety: Keep internal; anonymize client references when sharing insights.

1. "Summarize one thing that automation improved this week and one it complicated."

2. "Write a brief note to future-us summarizing this lesson and how to apply it."
3. "List three biases that appeared in our AI outputs and how we mitigated them."
4. "Generate a one-page recap of this project's 'aha moments' for internal training."
5. "Draft reflection questions for a team's quarterly AI review meeting."
6. "Design a five-question survey assessing how comfortable staff feel using AI ethically."
7. "List recurring small wins that indicate adoption success."
8. "Write a script for a five-minute peer teaching talk on responsible AI use."
9. "Create a template for documenting each prompt experiment with input, output, and lesson."
10. "Suggest metrics for evaluating AI impact on efficiency and learning."

Ethical and secure prompting checklist (One last time)

- Always anonymize private or proprietary info.
- Confirm your environment complies with your firm's data policy.
- Include a brief rationale ("why we're doing this").
- Ask the AI to show sources or confidence levels for factual claims.
- Read outputs through the *Reputation Test*: Would we sign our names to this?

Appendix C
Frameworks & Templates

Why this matters

Ideas only gain traction when they have a shape. And, particularly in the design professions, if you can visualize an idea, it's much more likely to 'stick.' The following frameworks and templates condense the key mental models from this book into tools you can print, adapt, and return to during projects and pursuits.

Each one has been pared down to its bare essentials. Use them as conversation starters, not corporate forms. Annotate them, modify them, and keep them close to where your team actually plans and reviews work.

---

1. The 3 Human Questions Framework

Purpose: To ensure that every client communication, whether it's been drafted by you or assisted by AI, reconnects to trust, relief, and relevance.

Question	What to Check For	Practical Test
Can I trust you?	Credibility, calm tone, proof of reliability	Does this sound experienced and steady—no hype, no ego?

Question	What to Check For	Practical Test
Will you make my life easier?	Clarity, structure, empathy for workload	Does this reduce anxiety or add new chores?
Do you get people like me?	Cultural fit, shared values, personalization	Could only *we* have written this for *this client*?

How to use:
Run every proposal, email, or presentation through a 90-second gut check. If "yes" isn't obvious across all three, revise before sending.

---

2. **The Human + Machine = Momentum Model**

Purpose: Distribute work intelligently between people and tools.

Layer	Human Strength	Machine Assist	Result
Seeing	Curiosity & context	Rapid pattern search	Insight
Deciding	Judgment & ethics	Option testing & prediction	Sound choices
Communicating	Empathy & tone	Drafting & format checks	Momentum with clarity

Use Case: Before assigning a task to AI, mark which layer it fits. The model clarifies who leads and who assists.

---

3. The Augmented-You Checklist

Purpose: Keep humans in charge while gaining speed.

1. Set clear **Intent** – "What problem am I solving?"
2. Provide **Context** – Audience, tone, and scope.
3. Choose safe **Inputs** – No client identifiers; anonymize data.
4. Maintain **Visibility** – Track what the assistant did and when.
5. Require **Human Review** – Edit, verify, approve.

Rule of thumb: Automation = delegation. Abdication = risk. Stay aware.

---

4. The 20-Minute Research Circuit

Purpose: Compress research into a focused rhythm of insight.

Step	**Duration**	**Action**
Define Intent	2 min	Phrase your research question clearly.
Collect Signals	8 min	Use AI to surface public, verifiable data.
Verify & Label	5 min	Confirm sources; Tag items as Signal / Pattern / Meaning.
Summarize & Share	5 min	Draft a one-page "Client Scan Sheet."

Making this a habit helps everyone learn what works, without adding more work.

---

5. The Purpose–Emotion–Reputation Judgment Checkpoints

Purpose: Ensure that you've reflected on what you're intending before you hit the "Send" button.

Checkpoint	Ask Yourself …	Why It Matters
Purpose	Why are we doing this at all?	Prevents "because we can" automation.
Emotion	What will this feel like to someone receiving it?	Keeps empathy in the loop.
Reputation	Would we be proud to own this publicly tomorrow?	Safeguards brand trust.

Run these in under one minute before approving any AI-assisted output. Build that all-important pause into the process.

---

6. Learning-Loop Template

Purpose: Turn everyday experiences, big and small, into practical lessons the firm can reuse.

When to use it: Right after any situation that taught you something worth remembering. The loop is a short review you run while the details are still fresh.

What to capture:

- A prompt or workflow that worked better than expected
- A tone or message that didn't land with a client
- A new pattern you noticed in how AI assisted (or confused) the work
- Any result you'd want to repeat, or avoid, next time

Four Questions:

1. **What happened?** – Summarize the event or result in plain language.
2. **What did it mean?** – Identify the insight or cause behind it.
3. **What will we change?** – Pick one small, concrete adjustment.
4. **Did it work?** – Check later to see if the change had an effect.

How to record it:

Capture answers in three short lines: *Situation* → *Observation* → *Adjustment*. Post them in your shared workspace under #LearningLoop so everyone can find and build on them.

---

7. Safe Experimentation Plan (Your AI Sandbox)

Purpose: Provide a risk-free zone for experimentation.

1. **Scope:** Internal only; no client/confidential data.
2. **Duration:** 10–15 minutes, once a week.
3. **Tasks:** Rewrite, summarize, brainstorm.
4. **Buddy:** Pair with another colleague for joint review.
5. **Reflection:** Note what surprised you or sparked curiosity.

Treat the space as practice, not production. Learning thrives where safety exists.

---

8. "Pause Moment" for weekly reflection

Purpose: Maintain ethical awareness and professional calm.

Each Friday, ask three team questions:

1. Where did automation actually help this week?
2. Where did we override a machine decision, and what did we learn?

3. Did our choices today reflect who we want to be known as?

Capture one insight, share one line publicly inside the firm. Ends the week with integrity, not fatigue.

---

9. Client-Care with signals & sentiment

Purpose: Translate digital data into empathy-driven action.

AI Sees →	**Humans Interpret →**	**Action**
Increase in reply time (slower response)	Possible workload spike or budget concern	Call to check in without mention of metrics.
Positive tone in feedback	Satisfaction + trust moment	Capture language for future case-study.
Silence longer than 30 days	Potential drift	Send personal note of appreciation, not template marketing.

Technology can surface the signals, but only people can turn them into understanding.

---

10. Ethical Disclosure Checklist (Clarity before hitting 'send')

Purpose: Normalize transparent communication about AI use.

- Disclosure – Plain statement of AI assistance where relevant.
- Confidentiality – No client/firm data in public systems.
- Human Review – Named reviewer confirmed accuracy.
- Tone – Sounds like us; aligns with our values.

- Record – Logged for training and accountability.

Keep this card next to proposal sign-off pages. It's the last gate before reputation.

Endnote

Frameworks and templates don't constrain creativity, they conserve it. By turning clarity, empathy, and responsibility into repeatable patterns, you free mental energy for insight and originality. Each template here is a reminder that machine speed works only when guided by human sense.

APPENDIX D
Glossary of terms and core concepts

Why this matters

Language shapes judgment. As technology accelerates, new terms enter practice faster than shared understanding can form. This glossary provides plain-language definitions tuned to the design and professional-services context. Use it as a reference sheet for teams learning to think, write, and decide alongside AI.

The glossary is not a dictionary; it's a field guide. As language changes, revisit these definitions with your team. Update them as your understanding deepens. The goal isn't to freeze meaning. It's to keep shared understanding moving forward with integrity.

A

AI Assistance / AI Assistant — A digital collaborator trained to process language or images that helps amplify a person's efficiency and insight. In this book, an Assistant is always supervised by a human reviewer with final accountability.

AI Governance Policy — A firmwide framework defining how AI tools are selected, used, and monitored to ensure ethical, secure, and transparent practice.

AI Sandbox — A safe, private environment where staff can practice using AI without risk to client confidentiality or project data; also called a *Safe Experimentation Zone.*

Augmented You — A model of professional growth where people use AI to extend capacity and thinking, not to offload responsibility.

C

Client Scan Sheet — A one-page summary of public information about a client's priorities, tone, and language patterns, created to align proposals or communication styles.

Confidence Corridor — The intentional pause before sharing a message or document: *Would we sign our name to this tomorrow?* A quick check that merges ethics with quality control.

Confidentiality by Default — A principle requiring that all client or firm data be treated as private and never entered into public systems unless explicitly secured.

E

Empathy Data — Qualitative signals such as tone, timing, and response patterns that help interpret human meaning behind numeric metrics. AI can surface these, but only people can interpret them responsibly.

Ethical Disclosure — A straightforward statement when AI assisted in content creation, showing transparency without apology or hype.

F

Framework vs. Template — A *framework* organizes thinking (why and when); a *template* standardizes doing (how and what). Frameworks help teams reason; templates help them execute.

H

Human + Machine = Momentum — The central equation of this book. Human context and machine speed working in alignment accelerate trustworthy progress.

Human Oversight — The practice of assigning a real person to review, edit, and approve any AI-assisted output before it reaches a client or the public.

L

Learning Loop — A four-question reflection habit (*What happened? What did it mean? What will we change? Did it work?*) used to capture small lessons before they fade.

Local Knowledge — The lived understanding of a community, market, or practice area; essential for judging whether AI-generated insights fit real conditions.

P

Prompt (Professional Prompt) — A written instruction to an AI Assistant built with the structure *Context* + *Action* + *Outcome.* The clarity and empathy of the prompt determines the quality of the result.

Purpose–Emotion–Reputation Checkpoints — Three self-checks before sharing AI-assisted work: *Why are we doing this? How will it feel to others? Would we be proud to own it publicly?*

S

Signals and Sentiment — Digital traces such as response time, tone, or frequency that may indicate a client's satisfaction or fatigue. Read as hints, never as proof.

Stewardship Mindset — An approach that sees each project, dataset, and relationship as something temporarily entrusted to our care—never exploited for short-term gain.

T

Three Human Questions — The universal test of trust and empathy in communication:

1. Can I trust you?
2. Will you make my life easier?
3. Do you get people like me?

Trust by Design — A professional principle that favors clarity, consistency, and care over speed or persuasion; the design of every process should lead naturally to credibility.

W

Weekly Pause Moment — A short Friday reflection where teams ask: *Where did automation help? Where did we override it? What did we learn?* Keeps ethics active without slowing momentum.

Further Learning

Professional practice evolves faster than printed references. Instead of relying on time-sensitive lists, return regularly to your own peer networks including SMPS chapters, AIA knowledge groups, ACEC's Resource Hub, and inter-firm roundtables to stay current with responsible, human-centered technology. Shared conversation will always outlast a static reading list.

ABOUT THE AUTHOR

Trained as an architect, David A. Stone has launched numerous companies and has helped countless architects, engineers, and marketers craft their strategies, shape their proposals and win a steady supply of profitable work. He is the author of more than 20 books, a much sought-after presenter and keynote speaker at gatherings around the world, and a trusted advisor to CEOs and Presidents of engineering and architecture firms in every region of the U.S.

Made in United States
Orlando, FL
04 November 2026

90218972R00213